FDR AND THE ENVIRONMENT

The World of the Roosevelts

General Editors: Arthur M. Schlesinger, Jr., William vanden Heuvel, and Douglas Brinkley
The Franklin and Eleanor Roosevelt Institute Series on The Roosevelt Era

FDR AND HIS CONTEMPORARIES
FOREIGN PERCEPTIONS OF
AN AMERICAN PRESIDENT
Edited by Cornelis A. van Minnen and
John F. Sears

NATO: THE FOUNDING OF
THE ATLANTIC ALLIANCE
AND THE INTEGRATION
OF EUROPE
Edited by Francis H. Heller and
John R. Gillingham

AMERICA UNBOUND
WORLD WAR II AND THE MAKING
OF A SUPERPOWER
Edited by Warren F. Kimball

THE ORIGINS OF U.S. NUCLEAR
STRATEGY, 1945–1953
Samuel R. Williamson, Jr. and
Steven L. Rearden

AMERICAN DIPLOMATS IN
THE NETHERLANDS, 1815–50
Cornelis A. van Minnen

EISENHOWER, KENNEDY,
AND THE UNITED STATES
OF EUROPE
Pascaline Winand

ALLIES AT WAR
THE SOVIET, AMERICAN,
AND BRITISH EXPERIENCE, 1939–1945
Edited by David Reynolds,
Warren F. Kimball, and A. O. Chubarian

THE ATLANTIC CHARTER
Edited by Douglas Brinkley and
David R. Facey-Crowther

PEARL HARBOR REVISITED
Edited by Robert W. Love, Jr.

FDR AND THE HOLOCAUST
Edited by Verne W. Newton

THE UNITED STATES AND
THE INTEGRATION OF EUROPE
LEGACIES OF THE POSTWAR ERA
Edited by Francis H. Heller and
John R. Gillingham

ADENAUER AND KENNEDY
A STUDY IN GERMAN-AMERICAN RELATIONS
Frank A. Mayer

THEODORE ROOSEVELT AND
THE BRITISH EMPIRE
A STUDY IN PRESIDENTIAL STATECRAFT
William N. Tilchin

TARIFFS, TRADE AND EUROPEAN
INTEGRATION, 1947–1957
FROM STUDY GROUP TO COMMON MARKET
Wendy Asbeek Brusse

SUMNER WELLES
FDR's GLOBAL STRATEGIST
A Biography by Benjamin Welles

THE NEW DEAL AND PUBLIC POLICY
Edited by Byron W. Daynes, William D.
Pederson, and Michael P. Riccards

WORLD WAR II IN EUROPE
Edited by Charles F. Brower

FDR AND THE U.S. NAVY
Edward J. Marolda

THE SECOND QUEBEC
CONFERENCE REVISITED
Edited by David B. Woolner

THEODORE ROOSEVELT,
THE U.S. NAVY, AND
THE SPANISH-AMERICAN WAR
Edited by Edward J. Marolda

FDR, THE VATICAN, AND THE
ROMAN CATHOLIC CHURCH IN
AMERICA, 1933–1945
Edited by David B. Woolner and
Richard G. Kurial

FDR AND
THE ENVIRONMENT

EDITED BY
HENRY L. HENDERSON
AND DAVID B. WOOLNER

FDR AND THE ENVIRONMENT
© Henry L. Henderson and David B. Woolner, 2005.

First published in 2005 by
PALGRAVE MACMILLAN™
175 Fifth Avenue, New York, N.Y. 10010 and
Houndmills, Basingstoke, Hampshire, England RG21 6XS
Companies and representatives throughout the world.

PALGRAVE MACMILLAN is the global academic imprint of the Palgrave Macmillan division of St. Martin's Press, LLC and of Palgrave Macmillan Ltd. Macmillan® is a registered trademark in the United States, United Kingdom and other countries. Palgrave is a registered trademark in the European Union and other countries.

ISBN 1–4039–6861–6

Library of Congress Cataloging-in-Publication Data

FDR and the environment / edited by Henry L. Henderson and David B. Woolner.
    p. cm. — (The world of the Roosevelts)
    Includes bibliographical references and index.
    ISBN 1–4039–6861–6 (alk. paper)
    1. Roosevelt, Franklin D. (Franklin Delano), 1882–1945—Views on conservation. 2. Environmental protection—United States—History—20th century. 3. Conservation of natural resources—United States—History—20th century. 4. United States—Politics and government—1933–1945. 5. New Deal, 1933–1939. I. Henderson, Henry L. II. Woolner, David B., 1955– III. Series.

E807.F335 2005
973.917′092—dc22   2004054139

A catalogue record for this book is available from the British Library.

Design by Newgen Imaging Systems (P) Ltd., Chennai, India.

First edition: March 2005

10 9 8 7 6 5 4 3 2 1

Printed in the United States of America.

Contents

ACKNOWLEDGMENTS

In the fall of 2002 the Franklin and Eleanor Roosevelt Institute, the Franklin D. Roosevelt Presidential Library and Museum, and Marist College hosted a remarkable conference entitled *Recovering the Environmental Legacy of FDR*. As the title suggests, the conference sought to reexamine the Progressive era conservation policies of the New Deal based on the principle that many of the programs and policies of this remarkable era stand at the root of modern environmentalism.

A conference of this intellectual scope and vigor could not have happened without the assistance of many individuals and the support of a number of key institutions. The inspiration for the conference came from Roosevelt Institute Cochairs Anna Eleanor Roosevelt, granddaughter of Franklin and Eleanor Roosevelt, and William J. vanden Heuvel. Without their vision and dedication to the legacy of FDR such a conference would not have been possible. We are also grateful to Dr. Cynthia Koch, the director of the Franklin D. Roosevelt Presidential Library and Museum and to the staffs of the FDR Library and Roosevelt Institute for their kind assistance. Special thanks must also go to Marist College president, Dr. Dennis Murray, for his strong support, as well as to Marist's dean of Liberal Arts and director of the Hudson River Valley Institute (HRVI) Dr. Thomas Wermuth, whose steadfast commitment to our efforts proved invaluable. Chris Pavlovski, the program director of HRVI, and Marist's Director Special Events, Valerie Hall also deserve recognition for their help in organizing the event.

Our appreciation goes out to Sarah Olson, the superintendent of the Roosevelt-Vanderbilt National Historic Site, and our friends at the National Park Service, who added a great deal to the proceedings. For providing us with a Hudson River Valley perspective of FDR's environmental legacy we would like to thank Deborah Meyer Dewan, Scenic Hudson's director of Riverfront Communities and a longtime environmental and community advocate; Cara Lee, the program director for the Nature Conservancy's Shawangunk Ridge Program; and Martin Shaffer, assistant professor of Political Science at Marist College, and a specialist in the politics of American Environmentalism. We are also grateful to Dr. Thomas Lynch and Dr. Richard Feldman of Marist's

Department of Environmental Sciences for their assistance, and to Dr. Ray Teichman, senior archivist at the FDR Library for his enlightening and humorous comments on the environmental holdings of the FDR Library. For his wise counsel on the intellectual dimensions of the conference and this book, our thanks go out as well to our dear friend and mentor, Roosevelt Historian, William Leuchtenburg.

As has been the case with a number of Roosevelt Institute conferences, we are extremely grateful to the late Jack Gartland and the Charlotte Cunneen Hackett Charitable Trust for their financial support of our efforts. Without this support, this conference would not have been possible. The conference organizers also wish to acknowledge the financial assistance of Marist College, the Hudson River Valley Institute, and the Franklin and Eleanor Roosevelt Institute.

Finally, we would like to extend our heartfelt thanks to participants of the conference who so graciously accepted our invitation to share their knowledge of this fascinating aspect of the Roosevelt era; and our appreciation to series editors Douglas Brinkley and Brendan O'Malley for their assistance in the final preparation of this book.

Henry L. Henderson and David B. Woolner
Chicago and Hyde Park

WILLIAM E. LEUCHTENBURG

For those of us who lived through the era of the New Deal, Franklin Delano Roosevelt (FDR) was the hero of the conservation movement, and multipurpose river valley development was our creed. We were thrilled by the dams in the Tennessee Valley and the Pacific Northwest, and proud that leaders all over the globe thought of the Tennessee Valley Authority (TVA) as the model for how to harness the energy of rivers to provide clean electric power and how to nourish "grassroots democracy." We acknowledged the pioneering efforts of FDR's distant cousin, Teddy, but we doubted that anyone ever had, or ever would, equal the initiatives of the New Deal: the Civilian Conservation Corps (CCC), the Soil Conservation Service, the shelterbelt, the Great Plains Committee, the wildlife refuges, and the national preserves from the Everglades and the Cape Hatteras National Seashore in the east to the Olympic National Park rain forest in the west.

Hence, it came as a shock when a later generation alleged that, with respect to the environment, FDR was not a hero but a villain. Yes, it is said, the New Deal did close the public domain lands to further entry, but it did so by permitting stockmen to exploit them. The TVA, which a British historian once marveled at as "the dazzling TVA," is now excoriated for fostering strip mining of coal, attempting to consign the snail darter to extinction, and fouling the air with sulphur dioxide. "Grassroots democracy," critics charge, means succumbing to the most powerful local interests. Instead of celebrating the dams we so much admired, environmentalists are destroying them to provide swift-moving streams for salmon and trout, just as well-intentioned New Deal public housing projects have been imploded.

These conflicting perceptions raise a series of questions. Some of the questions invite scrutiny of the record. Did Roosevelt knowingly jeopardize the environment in pursuit of other goals such as combating unemployment? Was FDR, with his utilitarian bent, insensitive to the need to keep the wilderness pristine? Other questions are exculpatory. Is it fair to blame FDR for policies adopted by TVA officials appointed by his Republican successors?

Is it reasonable to criticize the New Dealers for not grasping the nuances of environmentalist theories that did not emerge until a generation later? Still other questions encourage reflection. Has the criticism of the New Deal by modern-day environmentalists been too harsh? Have they failed to perceive continuity between Roosevelt's actions and the priorities of today? Richard N.L. Andrews, while noting the deficiencies of FDR's performance, has concluded, "The New Deal era left an unprecedented legacy of conservation achievements to American environmental policy. No other era in American history produced such an extraordinary record both of restoring and enhancing the environment, and of creating an improved sense of harmony between human communities and their environmental surroundings."

The Franklin and Eleanor Roosevelt Institute is to be congratulated for calling a conference to explore these questions and for making possible this important book. Both are unique. Never before has the performance of an administration with respect to the environment been appraised so diligently. Never before has such an examination been undertaken in the spotlight of contemporary issues—and at a time when the current administration is under fire for lacking FDR's vision. This book not only gives us a fresh view of one of the most significant features of the age of Roosevelt, but also informs our understanding of the directions we should pursue in the twenty-first century.

Introduction

Henry L. Henderson and
David B. Woolner

THE PRESIDENCY OF FRANKLIN D. ROOSEVELT (FDR) had a profound impact on the environment of the United States. Through the effective promotion of public works projects, energy initiatives, natural resource management, hydrological planning, and the reform of industrial and agricultural practices, FDR framed the terms, conditions, and power relations that formed the basis for the development of national environmental agencies, policies, and regulations that followed his presidency for years to come. Yet, as William Leuchtenburg has noted, much of FDR's impact on the environment remains unrecognized and obscured, due in large part to contemporary criticism of many of the New Deal's efforts to harness our nation's resources for social and economic benefit, and the tendency of those who admire FDR's environmental achievements to focus on his efforts as a conservationist in the tradition of Theodore Roosevelt (TR). Certainly, as a great lover of the land, and as an individual who often characterized himself as a "forester," FDR's association with his famous cousin is well deserved. But FDR's approach to nature was less romantic than TR's. Although FDR actively led the nation in expanding protection of natural areas, preserving the national patrimony and pioneering conservation policies, he saw the environment not as a realm set apart from humans, but as the field for human action, inextricably linked with the human community, economy, and system of values. It was this complex and pragmatic understanding of the relationship of nature and humanity—made manifest through such projects as the Tennessee Valley Authority (TVA), rural electrification, the Civilian Conservation Corps (CCC), and the Soil Conservation Service—that informed and motivated the New Deal, and which set it apart from previous conservation movements.

It is important to recall that FDR's presidency met two of the most dangerous challenges yet to threaten the nation: the Great Depression and World War II. It is often asserted by critics of environmental policies that the concerns of the environment are wholly secondary to national security and economic survival; that the environment can be entertained after economic and security issues are addressed. Franklin Roosevelt's legacy, established during extraordinary national economic and security emergencies, teaches otherwise. It is time to recognize that FDR's legacy must be read in important part as an environmental legacy, directly related to his leadership in economic and national security matters.

To gain a critical understanding of this rich and complex history, and to explore ways in which it might be applied to present and future environmental policy, the Franklin and Eleanor Roosevelt Institute (FERI), the Franklin D. Roosevelt Presidential Library, and Marist College hosted a two-day conference entitled *Recovering the Environmental Legacy of FDR* in the Fall of 2002. The conference brought together scholars, public policy-makers, environmental activists, and students from across the country and from FDR's own beloved Hudson River Valley for an intensive two-day examination of this important topic. The eleven papers that make up this book were drawn from this conference. Together they provide an important new perspective on FDR and the environment.

Divided into four parts, the book opens with a biographical look at FDR as steward of the land. Here, we explore the roots of his environmental ethic and philosophy, and the impact of his beliefs on his policies and practices as landowner, governor and president, or as he once put it, as "Forester-in-Chief." In "Grassroots Democracy: FDR and the Land," John F. Sears confirms that there is no better place to start than FDR's beloved estate in Hyde Park, New York, the land that gave him the sense of place from which he drew so much of his personal philosophy. Sears insists that we cannot fully understand FDR, unless we understand his relationship to "the land," and his firm belief—like Thomas Jefferson before him—that the land and the people who work it are intimately connected, and that it is impossible to restore one without restoring the other. In "The Complex Environmentalist," Brian Black takes Sears's arguments one step further by asserting that FDR was in fact an "environmentalist," and that the roots of the modern American environmental ethic can be found in the New Deal. Moreover, Black insists that much of the social, ecological, and cultural underpinnings of the New Deal's environmental policies reflected FDR's thinking, including the permanent shift in the placement of the federal government as the overseer of national resources policy. The final essay in the opening section comes from Neil Maher, who in "A Conflux of Desire and Need": Trees, Boy Scouts, and the Roots of Franklin Roosevelt's Civilian Conservation Corps, links FDR's decision to launch the CCC to the progressive era philosophy of Gifford Pinchot, and to FDR's personal involvement in the Boy Scouts of America. Out of this interesting mix came the very practical idea of combing the conservation of natural resources with the conservation of young men and the creation of one of the most successful and well-loved programs of the New Deal.

Having examined the roots of FDR's environmental ethic, part 2 focuses on FDR as conservationist in two key areas: wilderness and rural agricultural America. Paul Sutter begins this examination with his stimulating essay, "New Deal Conservation: A View from the Wilderness." Here, the creative tensions between the establishment of an organization dedicated to the creation and protection of a system of wilderness areas on America's public lands—the 1934 Wilderness Society—is juxtaposed with the New Deal impulse to open those lands to outdoor recreation. Sutter follows the careers of three of the society's founders—Aldo Leopold, Benton MacKaye, and Bob Marshall—to gain a greater understanding of the incongruities involved in both preserving and developing wild nature, and through this examination, helps us make sense of an aspect of this era that environmental historians have largely ignored. Sarah T. Phillips's work examines the 1920s origins of two programs that would become central to the New Deal attempt to restore what FDR called the "proper balance" between man and nature in agricultural America: rural electrification and land-use planning. In this pathbreaking study, Phillips details the tremendous early work that went on in both these areas under President Herbert Hoover, but due to Hoover's disinclination to use the Federal Government as an active instrument to execute these programs, both languished until FDR decided to embrace them wholeheartedly in his 1932 run for the White House. The result would be the first real attempt to develop nationwide sustainable agriculture in American history.

The idea that policies for resource use should be based on sound planning and scientific information, backed up by a body of environmental law, encompasses the third part of this book. Accordingly, this section opens with an essay by professor of law, A. Dan Tarlock, who argues that modern environmentalism, including its reliance on law and regulation as a means to extend the planning horizon of governmental mandates, owes much to the New Deal. Tarlock insists that it is time to correct the widely held view that the New Deal is "virtually a blank space in the history of modern environmentalism," by recognizing that the Roosevelt administration did far more than merely extend the work of progressive era conservation. The use of executive authority to protect and preserve our nation's natural resource heritage is the subject of John Leshy's essay on FDR's expansion of protections to natural resources, and how his creative use of authorities such as the Antiquities Act set important precedents for advancing protections to the environment in the Clinton Administration. Leshy acknowledges the rightful place of TR as the Act's founder, but he reminds us that FDR used the Act more often than any president, when measured by the number of monument proclamations. Leshy also describes how FDR reversed the trend in the federal government to divest nationally owned public domain lands, and as such, did as much if not more than any other president to preserve and protect some of our nation's most precious resources. Brian Black then returns with an essay that examines the establishment of the TVA as perhaps the most important early example of land management and environmental planning—a model that would reemerge in the 1960s. Closing this section is James R. Lyons, who argues in "FDR and Environmental Leadership," that "in thought, word, and deed, FDR and the cadre of environmental leaders who served with him . . . demonstrated a commitment to environmental protection that may be unparalleled in American presidencies." The importance of this fact is made all the more significant when Lyons reminds us that this work took place at a time when environmental degradation had destroyed the productivity of much of our nation's land.

The restorative nature of the New Deal is brought home in the final section of the book. Here, in an essay that takes its title from the theme of the conference that inspired this manuscript, Richard Andrews reminds us of how important it is to recover FDR's environmental legacy, to reclaim the past, and to recognize the three broad elements that the New Deal passed on to future generations: a legacy of specific environmental policies, programs, and institutions; a legacy of values and principles for environmental leadership and management; and a legacy of environmental results, including the many positive environmental consequences of initiatives intended to achieve other policy goals. Fittingly, the last essay of this book, "A New Deal for Nature—And Nature's People," by Roger Kennedy, looks toward the future as well as the past. Kennedy sees the bold, persistent experimentation of the New Deal as a model for our own time. FDR may have entered the White House in the chaotic circumstances of economic collapse and Depression, he argues, but what made him successful in response to these terrible calamities was his willingness and ability to respond instinctively, not theoretically, to the conditions around him. Equally important, FDR and the New Dealers understood what Kennedy calls "the peril of the irrevocable," because they had lived through the Dust Bowl and other natural disasters and had learned to listen to nature's voices. We would do well, he warns, to listen to nature in our own time, as we transgress our limits by overpopulating the dry western states' fire-prone canyons and hillsides. Kennedy, like Leshy, describes how FDR's legacy of progressive environmental leadership informed his own service in government as the Chief of the National Park Service in the Clinton Administration.

At the opening of the 2002 conference, FERI Cochair Anna Eleanor Roosevelt remarked, "FDR's dedication to improving the lives of his fellow citizens through conservation, planning, and the development of a comprehensive approach to the environment provides lessons that are still relevant for today's world." We agree, and it is our hope that this outstanding collection of essays will help us recover the environmental legacy of this remarkable leader and his era.

PART 1

FDR AS ENVIRONMENTALIST

CHAPTER 1

GRASSROOTS DEMOCRACY: FDR AND THE LAND

JOHN F. SEARS

NO AMERICAN PRESIDENT, NOT EVEN WASHINGTON OR JEFFERSON, has been more rooted in a particular place than Franklin Roosevelt (FDR) or drawn more of his substance as a leader from the land on which he was born and grew up.

FDR's remarks when he laid the cornerstone for the FDR Library in November 1939 reveal, in a humorous way, a great deal about the intimate connection he maintained to the land and to Hyde Park throughout his life:

> Half a century ago a small boy took especial delight in climbing an old tree, now unhappily gone, to pick and eat ripe sickle pears. That was about one hundred feet to the west of where I am standing now. And just to the north he used to lie flat between the strawberry rows and eat sun-warmed strawberries—the best in the world. In the spring of the year, in hip rubber boots, he sailed his first toy boats in the surface water formed by the melting snow. In the summer with his dogs he dug into woodchuck holes in this same field, and some of you are standing on top of those holes at this minute. Indeed, the descendants of those same woodchucks still inhabit this field and I hope that, under the auspices of the National Archivist, they will continue to do so for all time.[1]

As this passage reveals, FDR had a physical intimacy with the land; he experienced it in a tactile way. Although in 1939 he no longer stretched out on the ground to eat strawberries, he had a keen memory of that experience.

The land existed for FDR in time as well as in space. It had a history, and that history was organically connected to the present. The past was physically present in the land in the form of those woodchucks. And they can still be seen today along the entrance drive to the FDR Library.

FDR also had a strong sense of geography. That pear tree is "about one hundred feet to the west of where I am standing now." His memories are located in space. He liked to know the relationship of one place to another. His interest in stamps was, in part, an international extension of this local fascination with the geography of his childhood.

Because of this intimate connection to the land, its history, and its geography, it gave FDR great satisfaction, as he was laying the cornerstone of his library, that he was almost literally planting the records of his administration in a place that was bound up with his personal history.

But the land also had a much longer history that gave his act a universal meaning. He hoped, he said, that the library "will become an integral part of a country scene which the hand of man has not changed very greatly since the days of the Indians who dwelt here three hundred years ago. We know from simple deduction that these fields were cultivated by the first inhabitants of America—for the oak trees in these fields were striplings three centuries ago, and grew up in open fields as is proved to us by their wide spreading lower branches. Therefore, they grew in open spaces, and the only open spaces in Dutchess County were the cornfields of the Indians."[2] Many of those oak trees are still there today.

Another characteristic of FDR's relationship to the land is that he liked to read the markers of human history embedded in it, those features that, to the informed eye, reveal how human activity and natural processes have together shaped the land.

FDR's reference to the cornfields of the Indians and to the unchanging character and use of those fields over time suggests another point as well. Unlike the city, the countryside has a stability, a continuity that FDR believed America needed. It needed it especially in times of turmoil, such as the fall of 1939 when the world had just gone to war: "This is a peaceful countryside," he said, "and it seems appropriate in this time of strife that we should dedicate this Library to the spirit of peace—peace for the United States and soon, we hope, peace for the world itself."[3]

Hyde Park, then, was more than a source of personal identification for FDR; it was a model for the America he wanted—and even for the world. The America he envisioned was above all democratic, composed of people like the country "neighbors and friends" to whom he addressed his dedication remarks. It was not a country in which the state and its leaders, whether Fascist or Communist, were supreme. For FDR the history of the United States was not just the history of great men and great events, but of ordinary people. He wanted Americans of the future to come to the library to "gain a less superficial and more intimate and accurate view of the aspirations and purposes of all kinds of Americans."[4]

"Of the papers which will come to rest here," he said, "I personally attach less importance to the documents of those who have occupied high public or private office, than I do to the spontaneous letters which have come to me and my family and my associates from men, from women, and from children in every part of the United States, telling me of their conditions and problems, and giving me their opinions."[5]

What could be more "grassroots" than this: documents recording the way ordinary people experienced the history of their time coming to rest in these ancient cornfields among the woodchucks that FDR remembered from his childhood?

FDR's passion for discovering or creating organic connections between the past and the present is evident in the way he literally built the past into some of the Dutch Colonial stone buildings whose design and construction he oversaw. Top Cottage, the hilltop retreat FDR built on the back of his Hyde Park estate in 1938, was constructed of fieldstone from the walls on his property, stone that the early settlers of the area had extracted from the soil as they cleared the land.[6] The Rhinebeck Post Office was a copy of the Beekman House on River Road that FDR remembered from his youth. The original house had burned down, but much of the stone from the old Dutch house was incorporated into the front walls of the new building. In dedicating the Rhinebeck Post Office in 1939, FDR noted that connecting buildings to the land and its history was part of a national effort being made by the Procurement Division of the Treasury Department in designing new post offices throughout the nation: "we are trying to adapt the design to the historical background of the locality and to use, insofar as possible, the materials which are indigenous to the locality itself."[7]

The word "landscape" is rarely the right word to describe what FDR sees or thinks when he talks about the Hudson Valley or any other piece of countryside. He does not see a picture or painting when he looks at the land, although picturesque scenes no doubt had an appeal for him; he sees, instead, a set of relationships among human beings and between them and the natural environment. Looking at a landscape as a painting, as scenery, places the viewer outside it, like a visitor looking at a painting in an art gallery. FDR was always a participant, not an observer. To him the countryside wasn't scenery, it was land. Perhaps this was why FDR seems not to have been strongly attracted to wilderness. He regarded land as something to be used—not exploited but put to the best economic purpose and managed in a responsible manner. In the case of his own property, he cared about its productive capacity not its potential as a work of landscape art. The question was how to manage the land well, in what we would call today a sustainable way.

FDR's special interest was in forestry. When he put down "farmer" as his occupation, he meant "tree farmer." He began planting trees in 1912, the same year he became chairman of the Forestry Committee of the New York

State Senate, and planted 1,000–4,000 trees per year until 1929. He continued to acquire worn out and abandoned farm properties adjacent to his own
in order to accommodate his passion for forest planting and management.
Beginning in 1929 he retained Nelson Brown, a professor at the New York
State College of Forestry at Syracuse University, as his forester and together
they drew up a forest management plan for his estate. From that year
on his plans grew more ambitious. Between 1930 and 1945 he planted
20,000–55,000 trees per year, including a large number of Christmas trees.
Overall, Brown estimates, FDR planted over half a million trees on his Hyde
Park land.[8]

As he worked with Nelson Brown to expand and improve his own forests,
FDR drew on his knowledge and passion for forestry to institute forestry programs at the state level. As governor of New York during the onset of the
Depression, FDR initiated a program to put 10,000 young men to work in
New York's State forests planting trees, creating fire roads, and controlling
erosion. This became the model for the Civilian Conservation Corps (CCC),
which FDR began eagerly planning with Brown's help in December 1932
even before he took office as president. The establishment of the CCC in
March 1933 was one of the first acts of his new administration.

Brown believed that FDR contributed more to American forest conservation than Gifford Pinchot, Theodore Roosevelt, and the other pioneers of the
movement. His greatest contribution, according to Brown, was in making
the idea of larger and better-managed forests familiar to the average person.
He did this through various organizations: the CCC, which over the ten years
of its existence employed more than 2.5 million men in conservation work;
the Soil Conservation Service, which provided expertise and resources to
private landowners in order to encourage forest plantings and improvement
and other measures to control soil erosion; the Tennessee Valley Authority
(TVA), which also strongly encouraged and supported good forestry and soil
conservation practices; the preparation in 1933 of "A National Plan for
American Forestry" by the U.S. Forest Service; the Shelter Belt Program, an
ambitious program to plant trees in a 100-mile belt reaching from Texas to
the Canadian border in order to protect crop lands against prevailing winds
and drought (33,000 farmers participated in planting 217,000,000 trees
under this program); the Flood Control Act of 1936, which gave the federal
government a significant role in protecting watersheds; and the Norris Doxey
Farm Forestry Act of 1937, which made the expertise of foresters available to
farmers. He also took the lead in expanding the national forest system and
supporting forestry research. As important as any of the specific programs
that he championed, was his ability to explain the importance of these programs and of national forest management to the public at large through his
speeches and radio addresses. The intimate knowledge and love of forests that

he had acquired through years of experience on his land in Hyde Park, and the hours he had spent discussing tree farming with neighbors and other foresters, made him extraordinarily convincing in communicating to the average person the principles of good forestry and the importance of forests to the national welfare.[9]

FDR's forests provided a good part of his recreation when he was in Hyde Park. Although he could no longer walk or ride horseback in the woods as he had as a boy, FDR creatively employed his open, hand-controlled Ford to get close to the trees and particular spots in the forest he loved. He employed a crew to build new wood roads and maintain and improve old ones, thus creating a network of at least 20 miles of roads in what might be called his "handicapped accessible" forest. These roads were often very crude, just cleared tracks in the woods with steep grades, sharp turns, and washouts caused by heavy rains. It took a strong car, plus skill and daring to dodge trees, rocks, stumps, fallen logs, and sometimes mud holes along the way. FDR achieved an extraordinary degree of mobility and autonomy by pushing his car to its limits, sometimes even abandoning the roads altogether and driving across fields and over brush as high as the car itself. Nelson Brown reports that when FDR was in Hyde Park, he usually went out driving twice a day. "We would often drive alone from one to three hours or more, resting in a shady glade for 20 minutes or more to informally talk things over. Or sometimes we would just sit and rest." Brown believed that in this environment, where FDR was "both figuratively and literally remote" from the cares and pressures of the White House, he could completely relax.

"He loved to rest in his car in some peaceful shady nook in one of the far corners of his place, such as the little pond near the top of a hill above his cottage and swimming pool. Another was in the deep glades of the hemlock woods below his place. Still another was the oak forest below his hilltop cottage." He could identify all the different trees on his property and also the shrubs, wild flowers, and birds. "[H]e knew the trees on his place as well as any forester," Brown said. He and Brown would often discuss the history of the forests in the Hudson Valley from the time of the Indians to the present and the many uses to which trees had been put, providing lumber to build houses and barns, posts, railings and stakes for fences, piling for wharves along the river, crossties for the New York Central Railroad, and poles for telephone and telegraph lines.[10] FDR himself sold logs for crossties to the railroad and arranged with Central Hudson Gas and Electric to use hemlock poles cut on his own property to carry electric wires to Top Cottage.[11]

In 1942, after the outbreak of World War II, FDR arranged to sell red and white oak on his property for use in building patrol boats and other craft for the navy. FDR must have been thrilled to be able to contribute personally in this way to strengthening America's defenses for the war; then, after the

United States entered the war, to the war effort itself. Each stage of the process: from marking, cutting, skidding, trucking, and milling the logs to building the boats was recorded by a photographer for a photo essay on the project for life magazine. FDR benefited economically from the sale of these trees, but this transaction clearly meant much more to him than that. In a speech in West Virginia in 1944 in which he advocated planting trees on the state's deforested hillsides, he proudly referred to his sale of oaks both as an example of the economic benefits of forestry and as a contribution to the war effort: "And in this war, back home, I cut last year—and this is not very Christian—over four thousand dollars' worth net of oak trees, to make into submarine chasers and landing craft and other implements of war. And I am doing it again this year."[12] In harvesting these oaks, he was once more rooting his vision of America literally and symbolically in the land of Hyde Park, creating a vital connection between local resources and national strength, and incorporating the products of the land into things of service to the nation. Just as his forestry experiments became a model for some of his programs to combat the human and environmental problems of the Great Depression, the sale of his oak trees became an instrument to combat foreign foes.[13]

The relationships FDR perceives in the land are fundamentally economic, but not abstract. They are ultimately human relationships: relationships between farmers and the land, among farmers in a community, and between farmers and the city folk who buy their products. The land, the plants and animals that inhabit it, the people who own it, the people who work on it, the community in which it is located, the beneficiaries of the products it produces are all bound together. The people living on the land are also a resource. As a resource they too require good management, but not through force. There is an undeniably paternalistic attitude in FDR, but he expresses it as a desire to bring people together into a community, with himself, of course, as its leader.

One of FDR's closest neighbors was Moses Smith who lived at "Woodlawns," a farmhouse at the intersection of Creek Road and Route 9G opposite Val-Kill that was torn down in 1975. FDR bought Woodlawns in 1911, the first piece of land FDR owned on his own, and in 1920 rented it to Smith who remained a tenant farmer on the land until two years after FDR's death. FDR liked to stop at Woodlawns to talk with Mose, as he was called, about farming. "FDR used to come over to talk about his tree farm," his son remembered. "[H]e planted Norway spruce and Douglas fir all over what is now Val-Kill—and he and my father would talk about seeds; and thinning, and commiserate on good years and bad. He loved to feel he was using the land to good advantage."[14] Nelson Brown reported that "Mose was probably the most frank and outspoken close personal friend of FDR. He had no hesitancy in using language to which he was accustomed—and this wasn't always language that was commonly heard around official circles of the regular or summer White House." FDR used to stop along the road to chat with neighbors from behind the wheel of his hand-controlled Ford. Brown remembers one such occasion when a large group had gathered around FDR's car along Violet Avenue. After a while, FDR asked if Mose was there. Mose came forward from the outside of the circle, took his hat off and his pipe out of his mouth, and asked deferentially if there was anything he could do for the president. FDR told a joke and "gradually," Brown reports, "the conversation became more friendly and intimate and soon Mose replaced his hat, put his big old pipe in his mouth and warmed up to the occasion. Finally, he fairly stuck his finger into FDR's nose and said 'Look here, young feller, I want you to tell me what you are going to do down there in Washington about this war business.' " Later, when FDR and Brown were alone in the woods, FDR said, "Say, did you see how far Mose got his finger into my nose?"[15]

When reminiscing for the Dutchess County Historical Society about the origin of the Fireside Chats, FDR said that in preparing them he tried to imagine himself talking to the average bank depositor. "Perhaps my thoughts went back to this land of individual citizens whom I have known so well in Dutchess County all my life."[16] Moses Smith would have been prominent among those listeners he pictured in his mind.

From 1934 until 1941, the front lawn of Smith's house was the site of an annual "Homecoming" organized around Labor Day by the Roosevelt Home Club, a nonpartisan organization created in 1929 to provide political support for FDR.[17] When FDR spoke at the first "Homecoming" in 1934, he had just returned from a cross-country tour seeing firsthand the conditions in the Depression-ridden nation and the results of measures that his administration had undertaken to confront them. He noted particularly the problems in the West created by the settlers from the East and Middle West who had cultivated a great deal of land "that ought never to have been cultivated." Characteristically, he suggested to his audience that they were organically related to these Western farmers through kinship, the intertwining of ethnic origins, experience, and common history. As a boy, he remembered hearing about an area around Brown's Pond north of the town of Clinton that people called "Kansas." FDR and other local historians wondered about the origin of this name. They concluded that it dated back to 1850, when a railroad agent persuaded six or eight families in that area, probably on marginal land, to get on board an emigrant train in Poughkeepsie and move out to Kansas territory. In his travels west of the Mississippi, FDR said, he was often approached by people with relatives back in Dutchess County. "It rather thrills me to think how this country all ties in together in that we, and when you come right down to it everyone of us has, proudly, an enormous number of cousins—they may be distant cousins—living in all sort[s] of places in the

United States . . ." This web of cousinship unifying the nation was reinforced in FDR's mind by the blending of ethnic strains. Unlike Europe, where nations struggled with each other, in the United States, he noted, "we have, most of us, got half a dozen racial strains in us and yet here we are, all Americans." Finally, FDR cited Lord Brice's observation that Americans have a "trying-out system through the different States." They can experiment with solutions to an economic problem in one part of the country, see if it works, compare it to experiments in other parts of the country, and eventually work out a solution.[18] This, of course, was the method FDR was applying through the programs of the New Deal. It was also the approach FDR had been using on his farmland in Hyde Park.

Having built up this vision of interconnection between the different parts of the nation and of the nation as a large economic laboratory, FDR came to the lesson he had prepared his Hyde Park neighbors to understand. On the surface, Dutchess County looked fairly prosperous ("no drought, pretty good crops"), but he hoped that the Home Club would have more meetings to which they would invite speakers to come and

> tell the truth about conditions and about the methods that are being used to try to solve those conditions. The more we do that, the more we will realize that if a farm family is on the verge of starvation in North Dakota, we people in the town of Hyde Park are helping to pay to keep that family from actual starvation; if we have made mistakes in the settling of the country in the past, we in the town of Hyde Park have got to pay to correct those mistakes. In other words, that we have a definite stake, not merely the spiritual side of it, or the social side of it, or the patriotic side, but the actual financial side of it.[19]

So, like Hyde Park, FDR's model for American democracy, the United States was a network of neighbors tied together by the land, economic and family relationships, and a common experience with deep roots in the past. Kansas and North Dakota were not far off places of no concern to the residents of Hyde Park; "Kansas" was in Dutchess County, just north of the town of Clinton.

In this and other speeches, FDR expresses a deeply organic sense of the relationships between the land and the people dependent on it for a living and between the past, present, and future. In his address to the Green Pastures Rally in Charlotte, North Carolina in 1936, FDR remarked that a writer had recently said of him that he "reverts to terms of land and water in his approach to any great public problem." He had to admit that this was true and in the speech he not only does so, but also demonstrates how deeply embedded such imagery is in the Western literary tradition. The speech is a sermon on economics and his text, inspired by the name of the rally, is the two resonant lines from the twenty-third psalm: "He maketh me to lie down in green pastures;/He leadeth me beside the still waters." These lines, he says, convey an "idealized security of the body and the mind." But FDR gives them an actual and concrete meaning by contrasting the images they evoke with the economic and environmental disasters taking place in the American West during the Depression:

> Green pastures! Millions of our fellow Americans, with whom I have been associating in the past two weeks, out on the Great Plains of America, live with prayers and hopes for the fulfillment of what those words imply. Still waters! Millions of other Americans, with whom I have also associated of late, live with prayers and hopes either that the floods may be stilled—floods that bring with them destruction and disaster to fields and flocks, to homesteads and cities— or else they look for the Heaven-sent rains that will fill their wells, their ponds and their peaceful streams.

FDR saw the restoration of the land—the prevention of dust bowls and floods through soil conservation practices, the rotation of crops, the planting of trees—as intimately bound up with restoring the livelihoods of the people living on the land: "long before I went to Washington, I was convinced that the long road that leads to green pastures and still waters had to begin with reasonable prosperity. It seemed axiomatic to me that a cotton farmer who could get only five cents a pound for his crop could not be in a position properly to fertilize his land, or to terrace it or to rotate his crops. . . ."[20] The well-being of people and land was inseparable; without the restoration of America's rural economy the restoration and conservation of its natural resources could not be successful. FDR would understand perfectly the current dilemma of trying to persuade the people of developing nations to fully protect their environments before they have achieved economic well-being.

The vision of the relationship between the land and the American people with which FDR imbued his speeches at the Green Pastures Rally can be traced back to the experiences in Hyde Park that shaped his thinking, not just in his youth, but throughout his life. Today most of us don't live in or even come from communities resembling the Hyde Park of FDR's day. As a result, the nature and meaning of FDR's relationship to the people and land of Hyde Park may be lost on us. To tourists Hyde Park is a set of house museums and a presidential history museum in a scenic setting. To the residents of Hyde Park it is bedroom community or a business community serving its residents and visitors. For neither the tourists nor the residents is it the farming community it was in FDR's time, much less the living, organic set of human, economic, and historical relationships with the land that were continuous with FDR's sense of who he was and what America was or could be. For FDR Hyde Park had a history, a vital present in which he as Franklin D. Roosevelt, "farmer," had an important role to play quite aside from his role as president, and it had a future. He had a deep knowledge of the town's history, acquired not only through reading, but also through exploration on foot and horseback and later, by car, through reading the markers of the past in the land, and through long chats with his neighbors. That history was rich with many personal threads. The future of Hyde Park was also personal, something he intended to help shape and graft on the past. Nothing thrilled him more than to be a shaper of the land and its human history. He took this consciousness with him into his presidency and wove its themes into his speeches. FDR's strength as a leader grew, in part, out of his ability to extend his economic, social, and emotional connection to the land in Hyde Park to the nation as a whole.

NOTES

1. "Address at the Laying of the Cornerstone of the Franklin D. Roosevelt Library, Hyde Park, New York. November 19, 1939" in *The Presidential Papers of Franklin D. Roosevelt,* Vol. 8 (1939) (New York: Macmillan, 1941), 580.
2. *The Presidential Papers of Franklin D. Roosevelt,* Vol. 8, 580–81
3. Ibid., 581.
4. Ibid.
5. Ibid.
6. John G. Waite Associates, Architects, *The President as Architect: Franklin D. Roosevelt's Top Cottage* (Albany, New York: Mount Ida Press, 2001), 29–30.
7. "Address at the Dedication of the New Post Office in Rhinebeck, New York. May 1, 1939" in *The Presidential Papers of Franklin D. Roosevelt,* Vol. 8, 302–03. For FDR's involvement in the design of Hudson Valley post offices, see Bernice L. Thomas, *The Stamp of FDR: New Deal Post Offices in the Mid-Hudson Valley* (Fleischmanns, NY: Purple Mountain Press, 2002).
8. Nelson C. Brown, "Personal Reminiscences of F.D.R." Typescript, FDRL, Small Collections, Nelson Brown, Box 2, "Material Relating to Roosevelt's Hyde Park Estate, 1930–1951," 3, 6.
9. Brown, "Personal Reminiscences," 9–12.
10. Ibid., 30–31, 17–18.
11. Waite, *The President as Architect,* 47.
12. "Informal, Extemporaneous Remarks at Clarksburg, West Virgina. October 29, 1944" in *The Presidential Papers of Franklin D. Roosevelt,* Vol. 13 (1944–1945) (New York: Harper & Brothers, 1950), 381.
13. Correspondence documenting this and a subsequent sale of oak for ship building can be found in FDRL, Small Collections, Nelson Brown, Box 1, "Roosevelt, 1939–42," "FDR, Selective Cutting 1942," and "Logging Contract 1944."
14. As quoted by Helen Meserve in "The House that Became a Second Home to FDR," *Hyde Park Townsman,* September 7–8, 1983, B1. FDRL Miscellaneous Documents, "Arthur Plog."

15. Brown, "Personal Reminiscences," 15–16.
16. FDR to Helen W. Reynolds, October 30, 1933. FDRL, PPF #234. For a fuller discussion of FDR's relationship to his Hyde Park neighbors, see F. Kennon Moody, "Franklin D. Roosevelt and his Neighbors," Dsst., State University of New York–Albany, 1981. The author gratefully acknowledges Dr. Moody's research assistance in preparing this paper.
17. Meserve, "The House that Became a Second Home."
18. "Informal Extemporaneous Remarks of the President to the Roosevelt Home Club Members and their Families, Moses Smith Cottage, August 30, 1934." White House stenographer's transcript from shorthand notes. FDRL, FDR Master Speech File, #735, 3–7.
19. "Informal Extemporaneous Remarks of the President," 7–8.
20. "Your Life and Mine, though We Work in the Mill, the Office or the Store, Can Still be a Life in Green Pastures and beside the Still Waters—Address at the Green Pastures Rally, Charlotte, N.C., September 10, 1936" in *The Presidential Papers of Franklin D. Roosevelt,* Vol. 5 (1936) (New York: Random House, 1938), 342–45.

CHAPTER 2

THE COMPLEX ENVIRONMENTALIST: FRANKLIN D. ROOSEVELT AND THE ETHOS OF NEW DEAL CONSERVATION

BRIAN BLACK

WHEN THE NATION'S LEADER ASSURED CITIZENS GRIPPED by the difficulties of the Great Depression that they had "nothing to fear but fear itself," he needed to immediately manufacture substantiation.[1] The source of such optimism might be most readily found in the spirit of Americans' propensity for hard work and innovation. However, after discussing this fact, the leader made an extension that drew directly from his own passion and experience for the natural environment. "Nature still offers her bounty and human efforts have multiplied it," he continued. "Plenty is at our doorstep, but a generous use of it languishes in the very sight of the supply."

By the time Franklin Delano Roosevelt (FDR) made his presidential inaugural in 1933, his personality and life were managed in severe partitions.[2] While he appeared outwardly affable and jovial, he maintained such optimism only by confronting inner restrictions and limitations. He had spent the majority of the 1920s learning to physically live within a body wracked by polio; more importantly, his affliction had forced him to assess and redirect his own life and aspirations. In short, there were certain passions within his personality that FDR would never again be able to approach.[3] Before polio, "passionate" was a frequent adjective used to describe FDR; afterward, the most common adjective was "determined."

The complexity of this personal struggle and the psyche with which it was managed help to make FDR one of the most studied historical figures of American history.[4] Few historians, however, have sought to explore the continuity within the man before and after polio. In fact, intellectual convictions exist contiguously that overcome the partitions in these distinct periods of FDR's life. As a case study, this essay explores the relationship that FDR possessed with the natural environment. This was no simple intrigue; the natural environment's significance to human welfare composed one of FDR's most basic—almost intrinsic—convictions. His interest ran from the intellectual to the practical, including models of public policy and survival rates of specific tree species. Most importantly, FDR's environmental ethic grew from a Jeffersonian ideal that a connection to the land and hard work helped to make better people and better Americans.[5] By extension, though, FDR believed the use of nature needed to be orderly and sustainable. Humans were entrusted with a custodial responsibility to properly manage the natural resources around them.

Certainly, any exploration of personal ethics contains biographical significance to the life of this great leader, but in this examination our greatest concern is to highlight the "other" Roosevelt's crucial importance to the American environmental movement. Pursuing a better understanding of the importance of the natural environment in FDR's personal convictions allows scholars to better contextualize his actions as the nation's chief executive, including his inaugural address. In such a light, FDR becomes a seminal figure in the rise of American environmentalism, possibly even challenging the pathbreaking environmental record of his second cousin, Theodore.[6] While his inability to move about drastically limited FDR's ability to administer and interact with his natural surroundings, the largely immobile administrator devised the modes with which Americans interacted with and used nature. These policies made the New Deal a watershed in environmental history. At the core of his policies lay FDR's personal environmental ethic, a concept only perceivable in the continuity of his earlier and later life experience.

VITALITY AND THE "ENVIRONMENTAL" PERSONA

Scholars and general readers have long idealized the nineteenth-century environmentalists who swung the nation from a utilitarian view of the natural environment based largely in capitalism.[7] Individuals, including Henry David Thoreau, John Muir, and George Perkins Marsh, amaze contemporary readers with their ability to voice the opinion of an extreme minority during an age that blindly prioritized economic development.[8] In most instances, the naturalists who became fledgling environmentalists possessed a clear, physical connection to the natural environment. Often, these thinkers were identified with specific locales, including Thoreau's Walden and Muir's Sierras. Most often their thoughts derived from physical activity, farming, hiking, camping, or canoeing, that then established a connection between spirit, action, and place. During the Victorian era, the connection between environmental awareness and activity in the wilderness suggested that most such figures would be male.[9] A virility of active experience in the outdoors—even activities that may seem ironic to preservation such as hunting and fishing—became a crucial part of early environmentalism.

Through these tendencies, Americans shaped a personification of the environmental ideal. In his introduction to *The American Conservation Movement,* Stephen Fox explains why his history of early conservation became also a biography of the outdoor enthusiast and naturalist John Muir. Fox writes that during his own search for meaning in 1980s environmentalism, he found that Muir ". . . still seemed an active force in the movement. . . . When I looked at Muir more closely, I found that his religious ideology and the part he took in the movement set patterns for his successors in conservation."[10] In *Wilderness and the American Mind,* Roderick Nash emphasizes the impact of Gifford Pinchot and Theodore Roosevelt who took the spirit of Muir and began shifting it to active policy initiatives. Nash, however, points out that the "wild" spirit of early conservation remained intact, especially in the form of characters such as Joe Knowles whose nude disappearance into the wilderness in 1913 was widely covered by the news media.[11] While Nash is careful to draw out the culture and spirit of early environmentalism, *Conservation and the Gospel of Efficiency* by Samuel P. Hays emphasizes the political dimension of these developments. Hays accuses historians who have linked conservation directly to progressive ideology of concentrating ". . . on incidents which easily fit that viewpoint" and avoiding problems of other issues that fit less well. Hays's seminal study guides historians to look at conservation as a systematic intellectual shift. Conservation, he urges, is less about experience in wilderness and more about resource use becoming organized through scientific understanding.[12] More recently, scholars such as Richard Judd have complicated this picture by studying early land users—particularly farmers in New England—with ethical roots of the conservation ethic that lead to modern environmentalism.[13] The vast majority of early environmental thinkers were active on the land in one fashion or another. While Gifford Pinchot and Theodore Roosevelt (TR) first connected environmental concern to regulative policies, neither required that the persona of the "environmental ideal" lose its vitality of one-on-one experience with nature. Particularly in the case of TR, vitality compensated for a life of privilege.

TR, similar to many early environmentalists, knew nature through the manly experience of hunting, camping, and warfare. His interest in occupying the political spotlight inspired the stereotypes of early environmentalism, even organizations such as the Boy Scouts, Izzak Walton League, and Woodcraft Indians.[14] TR's larger-than-life image defined a new type of environmentalist, more often referred to as a conservationist. His robust efforts resonated to set contemporary standards—even allowing some observers to depict today's environmentalists as soft or passive in contrast.

Historian Richard White takes on some of these definitions of environmentalist in his essay "Are You an Environmentalist or Do You Work for a Living?: Work and Nature." Emphasizing contemporary definitions, White reproaches modern environmentalism for indifferently defining human labor as antithetical to environmental concern. Such a viewpoint restricts environmentalism to the likes of preservation efforts and less to efforts at conservation and management.[15] TR was one of the few politicians to contribute mightily to each branch of modern environmentalism.[16] No one doubted that TR was both an environmentalist and a willing laborer. His symbolic importance reached beyond the emerging environmental movement to encompass basic definitions of gender idealism. TR became a measuring stick of masculinity for all twentieth-century males. None, however, were likely to feel this as acutely as his young cousin, Franklin. By nearly every measure, though, FDR was no Theodore Roosevelt.

Even when his body allowed the vigor and physicality of outdoor exertion, Franklin generally selected a more intellectual approach. Unlike the famous exploits of TR, Franklin did not take his rifle and pursue confrontation with the greatest creatures in the world, nor did he go out of his way to seek out raw experiences in the wild, and he certainly led no one on horseback into dangerous, jungle combat. As the first executive to create national environmental policies of any weight, TR demonstrated his outdoor passions for the entire nation to see. He identified landmarks and wilderness areas to be given national designation and often visited them himself. His energy defined the meaning of the Progressive era in American politics and culture. The comparisons between FDR and this great American have consistently prohibited historians from accurately construing FDR's own significance in the history of environmentalism.[17] If one attempts to look at young FDR apart from these other, more traditional environmentalists, he finds that FDR's strong commitment to the natural environment fueled an ethic much more complex and applied than that of his famous conservation-minded cousin.

If one wishes to study FDR's relationship to his natural surroundings—his environmental ethic—the task is made even more difficult by his physical disability.[18] The major hurdle when studying FDR's life before 1921 is to be able to discern continuity where it, frankly, is obliterated by polio contracted at Campobello Island during the summer of 1921. Any interest in enjoying the out-of-doors is squeezed from the man's life in the 1920s to make room for the all-encompassing need to recover some use of his body. For the seven years following his illness, Franklin's obsessive personality drove him in his unwavering effort to rebuild a portion of his past mobility. Through these dual impacts— polio and his drive to recover from it—polio completely altered FDR's existence. Even when FDR returned to the passions of his pre-polio life in the 1930s, he, of course, remained almost entirely unable to interact with the natural environment due to his physical limitations. Therefore, scholars must look particularly closely for *any* interaction that he manufactures and multiply its significance commensurately.[19] FDR's record of action undoubtedly demonstrates a strong conservation ethic, thereby requiring historians to expand our stereotypes of the environmental ideal to include the immobile bureaucrat.

In order to recreate this record, one must seek out the continuity in Franklin's life. While most of his familial connections become strained or entirely crumble during his recovery period, FDR maintained a meaningful continuity in his life before and after polio: his family's Hyde Park estate.[20] In this setting, scholars might begin to reconstruct the environmental ethic of the thirty-second American president.

NATURAL INTRIGUE AT "THE PLACE"

The defining point of eastern New York is the broad, winding Hudson River. The river carried the first Europeans into the regions of Native occupation, including the Mohawk, Iroquois, and others. The currents of the river also aided early European farmers to move into the surrounding hills and lowlands as they practiced agriculture on a variety of scales. The river carried their crops to larger markets and helped to create New York City at its mouth. And the river then carried in the opposite direction emblems of urban culture into the countryside. Particularly, the river provided transportation and aesthetic sublimity to help make the valley, feeding it one of the first rural getaways for the evolving upper class of America's Victorian era.[21] Following the initial periods of European agricultural settlement, many small-landholders sold out en masse to aristocratic Americans who wished to construct sprawling estates on large parcels of land.

Roughly halfway between New York City and Albany, the state capital, lies Hyde Park. Within this small community, the Roosevelts owned a 1,200-acre plot rising steeply from the eastern bank of the Hudson. The family home, regal but conservative relative to nearby homes such as that of the Vanderbilts, topped the bluff and allowed majestic views of the river. Following its agrarian roots, Hyde Park had centered itself on providing comfort to the leisure society of New York City.

To call such a locale home while one grows through adolescence would seem a storybook fantasy. For young Franklin Roosevelt, born in 1882 to

James and Sara Roosevelt, his indulged upbringing helped to make Springwood into his personal, magical kingdom. Geoffrey C. Ward, biographer and historian, writes: "Franklin was brought up in splendid but lonely isolation on the Hudson . . . ."[22] Franklin's infrequent playmates were close relatives, children of neighboring country gentlemen, or children of employees who were encouraged to follow Franklin's wishes. Throughout his adult years, the Hudson Valley estate remained one of Roosevelt's most cherished devotions. From the start, the young Roosevelt's doting mother Sara raised him to be president, or at least a significant leader in American politics. Amidst such pressure during FDR's upbringing, Springwood served at once as oasis from public scrutiny and also the country setting of a life worthy of leading the nation through economic collapse and the world through war. From his earliest years through his death in 1945, FDR openly made every effort to associate himself with the Springwood Estate in Hyde Park.

From an early age, Franklin sought out a deep connection with Springwood. Years later Eleanor Roosevelt wrote: "[Franklin] knew every tree, every rock and stream on the place, and never forgot the people who had worked there when he was small."[23] When he was 13, Franklin collected one of each bird species on the farm. In the model of John J. Audubon, he killed one of each and stuffed them himself to maintain them for closer study.[24] Sara D. Roosevelt's diary offers some of the most concrete evidence that FDR loved nature from an early age and also records the constant activity and planning that she and Franklin orchestrated at the estate.[25] His interest in nature, trees, birds, and animals stemmed from his father, James Roosevelt, with whom Franklin rode the grounds of Hyde Park on horseback—having been given his first horse when he was six.[26]

By the age of five, Franklin also frequented the neighboring home of Archibald Rogers. The entire property had been carefully managed, with winding paths connecting fields and forests. The forests had been impeccably managed first by Rogers and then by hired supervisors. Impressive to the Roosevelts, the Rogers property generated beauty and profits from the worn soils of the Hudson Valley. On his 741-acre "Crumwald Farm," Archibald Rogers spent a great deal of effort to care for and preserve the stands of native trees, including experiments with reforestation and hiring forestry experts. His efforts attracted the attention of the U.S. Department of Agriculture and the division of Forestry made a complete survey of the Rogers property in 1906. Inspectors called "Crumwald Farm" one of the best examples of reforestation by a private owner in the state of New York.[27] When Franklin grew older, Rogers's ideal became his own wish for Springwood.[28]

For early indications of his interest in the nature of Springwood, one must look at Franklin's personal correspondence as well as the observations and writings of those nearest to him. In 1901, he wrote his mother from

Harvard: "You will be surprised to hear that I shall probably be home at Hyde Park for Sunday next. There are several reasons for this . . . also I want to get home again before all the trees are bare . . ."[29] While on his honeymoon, Franklin wrote to Sara about his conversations with other travelers: "I have had many long and interesting talks with Mr. Ferguson on forestry and with Mr. Foljambe and Mr. Kaye on farming and cattle raising, and the plans for Hyde Park now include not only a new house but a new farm, cattle, trees, etc . . ."

Around 1910, Franklin turned 30 and Sara began having him oversee more and more of the management of the estate. In the spring of 1911, she wrote that she had been: "worrying for a long time about our woods, and after a conversation with Mr. [Archibald] Rogers, who has studied the subject for years past with government experts, etc. and has worked hard himself, I have decided to go to work seriously this autumn and truly to clear all the dead wood out, a very large undertaking, and then to cut in the right way and *sell*. This will improve the woods, and we can begin to have young nurseries of trees." The estate represented an enduring link between Franklin and Sara. Eleanor later observed of Franklin: "He loved the land," she wrote, "and bought land contiguous to his mother's [194 acres in 1911], going straight back over the top of the hill, so that when he died he owned almost the area covering two of the original old water lots. He rented most of the farmland for just enough to cover the costs of taxes and insurance. His love of trees led him to start his Christmas tree plantations . . . ."[30]

Before he began his own projects for tree growing, Franklin researched the history of his family's estate. Franklin's environmental ethic began with one's need to know the site, both historically and ecologically. The land had produced prize-winning corn in the mid-1800s and had been farmed for over two hundred years. In 1910, the land was producing roughly half of what it had in 1840. He watched the gullies take shape and then carry the topsoil down to the Hudson. In a letter to Hendrik William van Loon, Franklin confided, "I can lime it, cross-plough it, manure it and treat it with every art known to science, but it has just plain run out."[31] Restoration of the soils became Franklin's mission for the estate.

In the model of his cousin Theodore, he was inspired to experience nature and learn its mysteries. Unlike his cousin, though, Franklin's interest in nature extended beyond the experiential and aesthetic to include a scientific interest in natural systems. In 1912, at the age of 29, Roosevelt created a record of his interest in the environment of his youth.[32] In the "Farm Journal" that FDR kept briefly from 1911 to 1917, one sees the construction of his commitment to serving as more than groundsfeeper of his family home; in the words of writer William Least-Heat Moon and others, Roosevelt sought to participate deeply in the "genius of a specific place."

Throughout his life, this particular locale remained the family's Hyde Park estate. He sought to learn its natural systems and to structure his own life around its patterns.

The "Farm Journal" contains no specific philosophical treatise. Instead, its detailed catalog of plantings and management reveals a level of attention rivaling the chronicle of an agronomist's forest science log. In the first entry of the "Farm Journal," Roosevelt uses the road as the organizing devise to then cordon off and list the contents of specific forested tracts. It was, after all, forests that interested Roosevelt. The hand-drawn maps of the estate provide readers with the clearest impression of Roosevelt's complete knowledge of Hyde Park. Roosevelt organized the property into lots, delineated either by number or name. He then kept organized records on each specific lot. The name for each lot derives from a natural feature of the site, including: River Wood Lot, Gravel Lot, North Farm Lot, South Farm Lot, Locust Pasture, Swamp Pasture, and so on.

In addition to recording each addition or subtraction to a lot, the Journal also included descriptions such as "Not worked" for the Northeast Wood Lot. For each area of the land, Roosevelt details the land—"swamp" "near railroad, must keep leaves raked"—and then the tree species that is most sensible to plant. The "Farm Journal" was obviously intended initially to be an accounting ledger of the human impact on the land. Roosevelt enumerates expenses on fertilizers and planting and eventually also lists any harvest and its value. This document, however, is entirely incomplete. While it demonstrates Roosevelt's intentions, the Journal also seems to have fallen prey to his increasingly busy life outside of Hyde Park. The entries are most complete during 1911 and 1912 before entirely trailing off by 1917.

THE SEARCH FOR CONTINUITY

Franklin's responsibility to Springwood and his mother evolved as he completed his education. Shortly after their marriage in 1905, Franklin and Eleanor faced a turning point that would determine a great deal about their future relationship. Of course, this revolved around what role Sara would play in Franklin's life, since before his marriage she had made almost every major decision for him. Following their marriage, the young couple initially refused to consider Springwood as a permanent home. They mentioned the estate's considerable upkeep as the reason; however, it seems likely that Sara's presence was also a factor. Sara was mortified that the family commitment to this place would not be passed on. Eleanor and Franklin maintained their own residence in New York City, but his involvement at Springwood grew annually. Within a few years, Franklin and Eleanor would come to spend a great deal of their time at Springwood, even completing an addition that included their own single bedrooms, with Eleanor's between those of Franklin and Sara.[33]

During the 1910s, Franklin's political aspirations began to be realized.[34] His meteoric rise ended temporarily with his unsuccessful run for vice president in 1920. This was the election that thrust Franklin on to the national scene.[35] One friend wrote Sara the following message after hearing Franklin speak on election eve:

> No one who heard your son . . . on Monday night need feel any concern for what the world would call his defeat. He is out of the reach of defeat. He is going his own way toward his own goal as steadily when voted down as when voted in. He is of a larger pattern than can be measured by any short sample of time. His *ideas* are as much beyond all power of defeat as he is himself. They are the *coming* ideas of the world . . . .[36]

In addition to his political potential, FDR had already begun to emerge as a "larger-than-life" figure. To many Americans, he appeared to be chosen for leadership and in possession of modern ideas with a "first-class temperament."[37]

With his temporary release from public office, Franklin could once again accompany his family on the annual extended vacation to Campobello Island in the Bay of Fundy in 1921. Before his political commitments, the family had gotten used to spending July, August, and part of September at their summer homes. As their son James describes, ". . . Campobello was his second home." While on the island he "inundated [the family] with fun. . . . Sometimes we felt we didn't have him at all, but when we did have him, life was as lively and as exciting as any kid could want it to be." Franklin was the center of things at Campobello, just as he was at Hyde Park, bursting with plans for picnics, always ready to sail or fish or camp.[38] In 1921, Franklin arrived worn down from overexertion in Washington; however, he seemed to drive himself even harder during his visit with his family. On August 10, Franklin contracted polio during exertion. For the first few months, doctors forbade him from even attempting to move fearing that his nervous system would be permanently damaged.[39] When the confusion settled, Franklin had permanently lost the use of his lower extremities.

Over the next seven years, the voracity with which Franklin approached his political hopes and all else channeled into his effort to gain enough control of his body to continue public life. He had been driven all his life to strive for leadership and power. Upon watching his cousin Theodore, Franklin had become steadfast in his aspiration to someday be president of the United States. These dreams were now all but extinguished; however, Franklin knew only to stand up to difficulty and to challenge it. In this quest, Franklin all but left his family and his former life. At times, Eleanor and Sara felt as if he had become someone altogether different. In fact, though, there remained continuity where it was possible. Strands of the life led before 1921 can still be identified in Franklin's life with polio. A primary interest remained Springwood and the science of forestry.

During his struggles with polio, FDR channeled energy into writing a history of Hyde Park that he compiled with a local historian in the mid-1920s. He was appointed local historian in 1926 and also worked specifically on a record of the family's estate. In this fascinating record, FDR obviously identifies the administration of the lands as his connection to the past occupants and particularly the men of his family. He makes specific and extended mention of the forests, including the forest along the River Road about which he writes: "It is one of the very few primeval forests on the river. It has never been lumbered and only live trees which had blown over have been cut. . . . " The document specifically contains a section titled: The Roosevelt Woods, subtitled "the dynamic story of conservation in a small but far-reaching package."[40] His local interests converged in a September 16, 1927, letter to the State Department of Highways that he titled "In Defense of Trees," which concerned the trees along the stretch of road connecting Springwood with the Vanderbilt estate:

> As you probably know, this particular stretch of road on the plateau on the top of Tellers Hill to Hyde Park Village has for many generations been noted for the magnificence of its trees, and this road is referred to in many books written about the Hudson River Valley. The trees were originally planted between 1750 and 1760, and many of the original trees are still standing. Others have been replaced from time to time, and it is a special matter of pride . . . that this stretch of road be kept up in its present fine condition. . . . I, with many others, will be very grateful if you can give special directions . . . that the trees are to be kept wholly free from any injury or damage which might result from the work.

Despite his physical limitations, Franklin found ways of expressing his passion for and interest in the natural environment. The most consistent expression of his interest in the cultural and ecological dynamics regarded Springwood. Geoffrey C. Ward writes:

> He was most consistent and effective when it came to conservation. And it was in this area, more than any other, that young Franklin Roosevelt sounded most like his older, steadier self. It was an interest to which he had been born, and in which he had been bred. Rides through the family forests alongside his father were among his earliest memories, Mr. James pointing out the special qualities of each variety of tree, making sure that his men preserved the finest old stands intact. And it had been the great cause of his cousin, Theodore.[41]

Whether intentional or not, the polio-stricken FDR would surround himself with people and policies that would continue many of the interests that he started in his early years at Springwood. Possibly no individual was more important in this fashion than Springwood's true "genius of the place."

During his youth, Franklin located some influences beyond those of the landed gentry of the Hudson Valley. Franklin was deeply influenced by individuals working at the Hyde Park estate. These were always professional relationships—typically unemotional and partly impersonal; however, such friendships still can be influential in one's upbringing. Though he strove to become the genius of Springwood, Franklin knew this honor lay with William A. Plog, who had come to work for Sara Roosevelt in 1897. Before being stricken by polio and particularly afterward, Franklin came to value the understanding and connectivity of Plog to this place for which he served as superintendent from 1897 to 1945. Plog was born in Poughkeepsie in 1869, where he also graduated from high school and married. Beginning as gardener for the Roosevelts, Plog would eventually supervise the gardeners and the wood-cutters, directed the harvesting of several tons of ice from the pond for summer drinks, look after the greenhouses and send cut roses to Sara at her town house in New York City. Generally, Franklin informed Plog of his wishes regarding general maintenance, tree harvest or planting, and so on.

From the White House, such directives were letters or an infrequent telegram. In letters addressed to "My Dear Mr. Franklin," Plog wrote the president frequently to consult about the mundane details of managing the property. Often such notes used the lot names that Franklin had created for his Farm Journal. These notes demonstrate a property owner still in charge of his "place," while he resided in Washington DC. Often, Plog's notes mentioned which shrubs and trees had been delivered and requested FDR's opinion on where they should be placed. In a 1937 letter to Plog, FDR stipulates the price to ask for Christmas trees. For little reason other than to demonstrate his own skill of observation, FDR ends the letter with the peculiar sentence: "As I remember it, the average in the lot is about six feet in height above the point we saw them off."[42]

THE MAKING OF ROOSEVELT FOREST

When Franklin's active oversight of the family estate began in 1910, he took little pleasure or particular interest in many of the tasks—these commonly referred back to Sara and eventually to Plog. The administrative role, however, did allow Franklin to develop the passion that he would carry on for the rest of his life: forestry and tree farming. Before the Roosevelts could follow Rogers's model for Springwood, the forest service had discontinued the program.[43] Franklin would have to initiate forest planning on the estate himself. His study of historical patterns and discussion with those versed in the science of forestry convinced Franklin that the soils of the Hudson valley had been exhausted by early agriculture. Similar to Pinchot, Franklin believed the resource could be replenished through better management. Specifically, organized efforts were needed to replenish the soil's nutrient base through the reintroduction of forest cover. From this point forward, Franklin listed his occupation as "tree grower" each year when voting in Hyde Park.[44]

In 1911, Franklin enlisted the assistance of the Syracuse University State College of Forestry to develop a reforestation program for the enlarged estate "in the hope that my grandchildren will be able to raise corn again—just one century from now."[45] This began his ongoing relationship with the school's dean, Nelson Brown, who described Franklin's activities as a forester as ". . . not just a passing fancy or plaything—it is very realistic and practical business endeavor."[46] When he was elected to the New York State Senate in 1910, Franklin began taking a more active role in the operation of the family estate, including Springwood's shift from crop farming to tree farming. He purchased adjoining farmland and in 1911 directed his farm crew to perform release cutting of hardwoods and to clear overgrown farmland for tree planting. In 1912, he placed his first order for seedlings from the New York State Conservation Commission. He would order more seedlings every year except for the years 1919–1923. From 1911 to 1945, Franklin oversaw the planting of thousands of trees on the estate. By 1945 he had supervised the planting of more than a half million trees covering 556 acres, or nearly one-half of the estate. In 1933 alone, FDR planted 36,000 trees at Hyde Park.[47]

Undoubtedly, Franklin's view of forestry was utilitarian. Harvests began in 1912 and continued until his death. Prime lumber from the estate was used for navy ships during World War I while Franklin served as assistant secretary. He also continued his efforts to expand regional interest in reforestation. Franklin's personal forestry always took the guise of an effort to educate regional farmers on the practice of reforestation. In 1914, he served as vice president of the State Forestry Association of New York. This effort grew more intense after his paralysis, when observers determined that he had adopted a "missionary state toward forestry." In a 1922 letter, Franklin proposed that a group of wealthy associates should form a company to practice and promote "scientific forest management."[48] In 1923, he promoted reforestation by offering to assist the local granges in efforts to secure seedlings. He also offered to make joint orders with the New York Conservation Commission. When their order did not arrive until after the deadline, Franklin's letter urged the supplier as follows: "I hope that the commission will approve of this order as I feel it is an entering wedge to get a large number of people interested in the planting of trees. . . ."[49]

Throughout the 1920s, Franklin urged lumber corporations such as the Pouvail Smith Corporation in Poughkeepsie to consider his forest for any needs for special sizes of white oak, black oak, locust, or hemlock.[50] Obviously,

Franklin's confidence in his knowledge and efforts grew each year. In 1926, his order to the commission revealed his own favoritism by requesting that they make tulip poplars available. He also related the following story:

> I hope you will forgive me for bombarding you with suggestions, but in driving around the Hudson Valley country this summer, I have been impressed with the lack of knowledge on behalf of farm wood lot owners as to which trees to retain and which to cut out. For instance, I educated one man as to the value of white oak. He was about to cut off clean a lot containing principally white oak, black oak, swamp oak and rock oak, and at my suggestion, he is cutting out everything except the white oak and will leave them for a few years in order that they may seed the vacant spaces.[51]

In fact, Franklin's growing obsession with the tulip poplar remained unsatisfied. His records show that he and Brown then began contacting commercial nurseries to purchase the seedlings.

In New York, Franklin worked hard to interest regional farmers in his ideas. In 1924 he wrote, "I had planned to put these trees into my own transplanting bed for one year or two years before distributing them to the 4 or 5 people in the Grange who wanted to try them out as I think they will take more interest in the trees if they look a little bigger when they are put in and will thereby be induced to keep them free from brush."[52] The return correspondence from the Conservation Commission urged Franklin that such interest-generation was helpful, but ". . . the handling of the farm wood-lot is of as much importance as is the reforesting of the idle lands." To this end, the commission requested that Franklin allow his farm to be set up as a county demonstration area.[53] An incredibly detailed letter from Newton Armstrong, Cornell University Arborist, demonstrates that Franklin's knowledge had become well respected. He wrote: "I have taken a great deal of your time in writing so long a letter but I know your great interest in all problems of this kind and have thought many times I would like to have an opportunity to discuss the question with you, not in relation to my little organization, but in its relation to the trees of the State of NY and, of course, the rest of the nation."[54]

While Springwood represented the core environment in Franklin's life, during the 1920s his efforts to rebuild his own body had taken his focus to other locations as well. At least in his mind, Franklin's physical recovery demanded that he frequent the natural spring baths located in Warm Springs, Georgia.[55] When the resort located at the spas ran into financial difficulty in 1926, Franklin's determination to recover fed a personal effort to purchase the property. During this period, Franklin had largely left his family behind. He spent much of his time in the American South and only brief periods with Eleanor and their family. Eleanor's correspondences regarding the purchase of the Warm Springs resort demonstrate the family's confusion and frustration over how best to help Franklin.[56] The Warm Springs purchase took on additional urgency to Franklin because he felt he was making significant progress in his recovery. Of course, such feelings boosted his own self-confidence and helped his demeanor. At the very least, while at Warm Springs Franklin made tremendous strides in his ability to move and manage his paralyzed body. For the purposes of this analysis, it is important to note that shortly after the purchase of the resort, Franklin began carrying out a forest development program at this site as well in 1927.[57] The interest and passion in forest conservation had become so intrinsic to his nature that Franklin would not overlook such an opportunity.

As the forests of Springwood became more complex, Brown took an increasingly active role. In 1930, Brown determined that the estate would act as a demonstrational forest. To carry out this process, he hired a recent Syracuse graduate, Irving Isenberg, to complete the forest plan that the Roosevelts had desired as early as 1905. In his correspondence to Franklin, Brown stipulated that the experiment station status "can be done without any definite commitment by you except to continue the excellent plan of management which you have apparently been following out in the past. It will be intended to improve the woods and to handle it as a commercial forestry operation."[58] The status must have pleased Roosevelt greatly—not to mention the commendation of his own efforts.

In his plan, Isenberg divided the property into 15 management units—14 were to be harvested when the market improved. The fifteenth unit, near the home, was regarded as old-growth or virgin woods and recommended for preservation. Isenberg wrote, "The stand should remain untouched. Do not remove even dead trees. Do not build new roads. Thus it will be preserved just as nature has treated it."[59] This stand of trees had long impressed Franklin. In his history of the estate, which was quoted above, he elaborated that this particular section of forest represented almost the only remnant of the Hudson Valley's natural ecology. The great trees linked his family to the past of the region in a way that Franklin relished.

Throughout the turbulent 1920s and into his presidency, the management of his forest interests seemed to provide FDR with the only consistent escape. After he became president, his continued oversight of the forest seems to have helped divert the stress of managing the nation through the Great Depression and then the world war. Only a few days before his death on March 24, 1945, he sent the following memorandum to Plog: "The chestnut trees are to go as fill-ins among the trees which did not do so well just east of the gravel bank in the field southeast of Linaka's house. I shall be delighted to have a few pear trees if Dr. Baruch has them. I will be back tomorrow. . . . Will you let Dr. Baruch's people know about the trees?"

"PRACTICE" IN STATE POLITICS

While Franklin hoped the model of his forests at Hyde Park and even in Warm Springs would inspire others, he also set out to create policy that could guide general land-use toward his ethic. His emphasis prior to 1933, of course, was the State of New York. In 1911 the newly elected State Senator Franklin Roosevelt was appointed to the chairmanship of the Senate Forest, Fish, and Game Commission, which was the direct forerunner of the State Conservation Commission. In 1913 he was also appointed the first chairman of the newly created State Senate Conservation Committee. The forestry work continued at Hyde Park from 1911 to 1916 and was likely a product of experience and knowledge gained while serving in the New York State Senate. The New York State Forestry Association was founded in 1913 and Franklin served as its vice president in 1915 and 1916 from his post in Washington, DC as assistant secretary of the navy.

When first elected to the New York State Senate, Franklin sponsored legislation to radically extend forestry legislation by regulating cutting on private lands. To gather support for his efforts, he invited Gifford Pinchot to Albany to address his committee. Pinchot spoke of a coming "timber famine" and urged the careful regulation of cutting on private lands. Eleanor recalled this speech years later:

> Mr. Pinchot was waging a rather lonely fight then and few people paid much attention to his warnings. However, my husband was tremendously impressed and he began at once to replant trees on his own land in Dutchess County, NY. He began to look wherever he went for soil erosion and he taught me to be conscious of this wasting of our land, too. The curious thing which is not always remembered is the fact that where land is wastefully used and becomes unprofitable, the people go to waste too. Good land and good people go hand in hand.[60]

Franklin undoubtedly hoped that Pinchot's discussion would help the Roosevelt–Jones Bill, which proposed to formalize and regulate forestry regulation and cutting on private lands. Such ideas, however, proved too progressive for New York state politics.[61] Of course, this could change if Franklin's ideas controlled the state's government. In 1928, Franklin ran for governor and won; he also won reelection in 1930.

As governor in 1929, FDR designed a rural program that enlisted the support of the state's farmers and foresters, together with the national forestry leadership. He urged preservationists to secure funds to expand the state's forests and politically buried proposals by upstate power companies to acquire leases on forest preserves in order to create reservoirs for generating hydroelectric power. In 1931 he made two speeches to support an amendment to the state constitution to allow the purchase of abandoned and submarginal farmland. Upon the voters' approval of the measure, an article in the *Journal of Forestry* said: "The present governor, so definitely a believer in forestry as to spend his own money in its practice on his personal estate, enthusiastically supported the measure. . . . As the nation's largest reforestation project, larger even than the federal government, much interest will be shown in its progress. . . . New York's enlarged forestry program is now an assured fact. No longer a political issue nor a subject to be campaigned for, it is now a matter requiring only technical accomplishment."[62] The suggestion here is that such policy grew from deep within the governor—out of his most basic commitments and ethics.

On this subject (as well as many others), Henry Morgenthau, Jr. helped Franklin shape his ideas. The two men had grown up together and now treated the governor's state experience as practice for the national stage. Morgenthau helped to organize the Governor's Agricultural Advisory Commission, which helped put into practice many of their ideas for rural New York. Historian John Morton Blum writes: "Both were the sons of wealthy fathers, both were comfortable, both enjoyed a degree of easy elegance, yet neither really cared for the toys and games of the adult rich. Both loved the land, the trees, the soil itself, all the things that conservation was intended to protect and develop."[63] Morgenthau aided Franklin most in taking idealistic concepts and making them into working plans. In the case of New York, the year after Franklin's first election as governor the state purchased 173,681 acres for forest expansion.

As Franklin neared a 1932 run for the U.S. presidential nomination, Alfred Smith began railing against many of the decisions that his political successor had made for New York. Smith focused directly on the proposed amendment to the state constitution that Morganthau had introduced for the Conservation Commission. The bill called for $20 million for reforestation in certain areas outside the Adirondack forest. Morganthau told one group: "The human species requires for a proper habitat something more than farming land, factory locations and urban areas. We need . . . common public playgrounds of wide extent to give us a place fit to live in and to live a healthy, balanced life. The waste spaces of today, the uncultivated areas which we plan to reforest and the existing woodlands form such a natural playground."[64] Contrary to Smith's urging, New York voters approved the amendment in 1931. New York then embarked upon a larger forestry operation than any state had previously attempted, including 10,000 men employed in 1,932 projects to reforest portions of the state. Such efforts, of course, foreshadow many New Deal efforts, particularly the Civilian Conservation Corps (CCC).

FDR made it clear in 1931 that he intended to function as a national leader on the issue of land conservation and land recovery. He addressed the

National Conference of Governors in the address titled, "Acres Fit and Unfit." Throughout the nation, he scolded, hundreds of farmers are cultivating land under impossible conditions. Their land simply cannot support agriculture and they have not been trained on how to better manage it. "They are slowly breaking their hearts," he concluded, "their health and their pocketbooks against a stone wall of impossibilities." Such land must be turned to the growing of trees, not produce. Similar to Pinchot, Roosevelt believed reforestation must be the obligation of the state and the nation.[65]

His passion for forest management and larger ideas of resource conservation must be viewed in the context of Franklin's life experience. When polio derailed so many of Franklin's plans and aspirations, he could still emphasize certain passions. Regarding conservation, Franklin became a different type of leader. Instead of virility and wilderness experience, he emphasized scientific management and ecological understanding. One can surmise that frustration at his own immobility fueled his efforts to create a new expression for his conservation ethic: that of administrator. Franklin's actions as an environmental executive enabled others to fulfill his own desires. In essence, American land-use becomes at least in part a vicarious expression of the early choices Franklin made at Springwood. FDR becomes a personification not of the wild land-tromping, passion-driven preservationist, but of the methodical, rational planner using modern design to implement his own aspirations for the entire nation. In the end, historians find an uncomfortable position: it is our stereotypes that require adjustment; the record leaves little doubt regarding FDR's convictions to conservation and the natural environment. He is merely a new type of environmentalist.

NEW DEAL NATURE

Following his 1933 election to the presidency, accounts of FDR often referred to him as the "Forester in Chief." It is likely, nothing could have pleased the man more. In his campaign, FDR promised to make forestry and conservation a national concern when he pledged a million men to work in a national "reforestation program." FDR's campaign accused the existing administration of a "narrow and negative attitude" toward forestry as merely "planting trees, whereas reforestation in the broad sense, with collateral work on forest protection, roads and trails, lookout towers, telephone lines, etc., would provide work for a very large number of men."[66] Historian Thomas Patton writes, "President Hoover's administration appeared ill-informed, unimaginative, and opposed to an effort that . . . many Americans considered a national imperative. . . ."[67] FDR brought up these deficiencies as well as setting out a course similar to what he had performed at Springwood and in New York state politics.

Ideologically, FDR's track record demonstrates conservation's importance to him; however, the vital image associated with this interest also gave the paralyzed man an inferred activity or connection with the outdoors. FDR fostered this image as well as his connections with his forest projects throughout his presidency. Upon FDR's request, Plog and Brown each maintained the estate and reported to the owner in Washington, DC. No matter the mundane detail, these letters merited immediate attention from the president. In 1942, for instance, Grace Tully sent Plog a telegram stating, "The President would like to know when the planting of the new tree crop starts. He might like to go home that weekend."[68] As one pores through the collected correspondence of the wartime president, it is jarring to see that on December 12, 1941 FDR wrote Plog that he hopes "the cutting and sale of the [Christmas] trees is going along well." He went on to request that Plog send two trees to family friends. Five days after the Japanese attack on Pearl Harbor and four days after declaring December 7 a day "that would live in infamy" and entering the United States into World War II, the tree farmer still concerned himself with his forest. Such letters recur throughout the war years and offer a stunning view into the priorities of the man who led the country through World War II. Throughout his presidency, FDR referred to his profession as "tree farmer" and it was Brown who allowed this claim to remain more true than hyperbolic. Additionally, FDR had Plog oversee the construction of roads on which his specially modeled vehicle could pass through Roosevelt Forests. FDR enjoyed the freedom of driving his car anywhere; however, touring the grounds and forests that brought him such pride was the highpoint of returning to the estate from Washington.

While such activity may not signify interaction with nature akin to hiking and camping, FDR's incapacity demands special consideration from historians. Due to his lack of mobility, his vigor for overseeing and knowing every detail of the forests represents an unwavering commitment to the careful use of natural resources. His ethic, obviously, grew out of a deep conviction in the necessity of human interaction with the natural environment. Through the expression of these efforts earlier in his life, one can see that FDR clearly was a vital and virile presence in the natural environment that he knew. His bureaucratic nature gained added emphasis when he was suddenly struck incapable of continuing many portions of his relationship with the natural environment. More than any public figure, FDR's political role increased the proximity of the American people to the natural environment and the understanding of its complex, scientific dynamics.

These complexities helped to define New Deal patterns of natural resource management. By infusing science and planning into policy making, the New Deal took the environmental impulse of the early conservationists such as TR and added a dose of modernity.[69] This version, of course, included the integration of human solutions—technology—into the natural environment in order to engineer its use. This is an environmental ethic based in human management, whether the laboratory is the Roosevelt Woods, Southern Plains, or a specific river valley. Each New Deal policy is based in the same priority that drove Franklin's efforts at Springwood.

FDR made certain that New Deal land-use emphasized minimizing waste. In the inaugural address that was quoted above, the new president urged Americans that they still had great wealth available, if the management of such resources were conducted with more care. In this speech, FDR's first job was to commodify the landscape in an entirely new fashion: he tied economic waste to mismanaged natural resources. Policies were intended to cut down on wasteful practices—whether those of nature or of humans—while putting Americans back to work. Efforts of agencies such as the CCC are often described by historians as little more than work projects; yet such a perspective overlooks the ethos behind them—a conviction emanating from the New Deal's oval office. In a very Jeffersonian sense, FDR believed labor on the land was a virtuous enterprise. While the national need was simply for employment, "works" projects provided an excellent opportunity to invigorate the minds and bodies of Americans—particularly young men.[70] The dynamic effort to rationalize resources in order to create jobs is yet another example of modernist thought being put into practice.

In each New Deal agency, the designed landscape became an enduring symbol of modernism. Of course, American landscapes had been subjected to conscious design for many years. The managed landscape became an entity that existed between nature and civilized, human space—depending on the details of its alteration. The details of each site derive from the ethos of the creator or conceiver. While the rustic landscape design of Frederick Law Olmsted and others often satisfy observers as a successful example of "the middle landscape" created from a marriage of civilization and nature, the modern version of the New Deal rarely receives similar consideration.[71] One reason for this slight is the New Deal ethos, which derived more from science than aesthetics. The sensibility to monitor and regulate natural resource use, therefore to conserve resources in some fashion, is an impulse claimed as the beginning of the environmental movement. Samuel P. Hays writes:

> The broader significance of the conservation movement stemmed from the role it played in the transformation of a decentralized, nontechnical, loosely organized society, where waste and inefficiency ran rampant, into a highly organized, technical, and centrally planned and directed social organization which could meet a complex world with efficiency and purpose.[72]

The idea of organizing and systematizing the natural environment—in other words, to create the "modern"—was not necessarily an effort to extract the natural environment from it. New categories of understanding, particularly ecology, literally added entirely new fashions with which conservationists could organize resources and add a new depth and understanding to the human–nature relationship. The infusion of ecological considerations into landscape design would critically alter the ethos behind the work of planners that, in many ways, fed the creation of a "middle ground" more faithful to natural systems.[73]

Careful management, of course, was based in a scientific understanding of the natural world. During this period, ecologist Frederic Clements and others furthered a line of ecological thinking based on the idea of the climax community, which held that nature should be used as a guide and the natural vegetative forms in certain regions and climates should be respected instead of altered.[74] Historian Donald Worster has written that during this period, some "reconciliation between the ecological and human patterns of evolution had to be devised, some accommodation reached between faith in nature's ways and sympathy for human ambitions."[75] During the 1930s, and largely through New Deal policies, Worster writes, came "the birth in the public consciousness of a new conservation philosophy, one more responsive to principles of scientific ecology."[76] While scientific planning need not involve aesthetics, New Deal land designers demanded visual appeal to symbolize a new ideal landscape and, in later years, to stimulate recreation. Worster writes that in the 1930s, ". . . conservation began to move toward a more inclusive, coordinated, ecological perspective."[77] Instead of specialized abstraction, he argues that "holistic values everywhere challenged private, atomistic ways of thinking, and the atmosphere of depression also encouraged an unwonted willingness to subordinate economic criteria to broader standards of value, including ecological integrity."

The implementation of such ideas in federal projects reflects a watershed change; however, New Deal agencies also set out to educate the general public regarding these changing ideas of nature and the human's role. In other words, FDR's government helped to construct a new environmental ethic for the American public. Hays writes:

> The deepest significance of the conservation movement, however, lay in its political implications. . . . The crux of the gospel of efficiency lay in a rational and scientific method of making basic technological decisions through a single, central authority.[78]

The new, modernist paradigm unabashedly gravitated toward technical solutions devised by experts; however it did not always require a shift from the "gospel of efficiency" reliance on a central authority. While the construction of this ideal would be construed by expert technicians in the sciences and landscape design, its implementation required a centralization and coordination of decision-making that was at first antithetical to the principles of American free enterprise.[79] The conservation impulse would fully combine with modernism in the person of FDR and in the projects of the New Deal.

PLANNING FUNCTIONAL NATURE

Other essays in this volume focus on specific efforts of New Deal conservation, so this section only emphasizes the general ethic that governed all such agencies. Whether administering forest, soil, wildlife, or water resources, FDR's policies grew from a general ethic of maintenance of existing resources without further depletion. Ecology was the main instrument for such management, with policies including the fledgling concept of sustainability. While FDR's actions resulted in the establishment of temporary agencies and actions, permanent shifts were also made in the overall placement of the federal government as the overseer of national resources policy.

The Soil Erosion Service, for instance, was established in 1933 to demonstrate the practical possibilities of curbing erosion and planning a landscape that could better weather the windstorms of the mid-1930s. These activities, of course, took their most significant form in the Taylor Grazing Act of 1934 and the construction of massive shelterbelts throughout the western U.S. The massive soil demonstration effort of the New Deal marked the first time in American history that the government had so aggressively sought to attack the problem of erosion. By 1934, 24 demonstration sites had been established.[80] Other sites would soon involve watershed projects as well. Often these sites were located on federally controlled land. By 1937, these efforts would result in the establishment of soil conservation districts throughout the nation. By 1940, laws established such districts in 38 states.[81]

Similarly, forest planning allowed FDR to fulfill social and economic objectives through the application of scientific conservation. In 1933, Senator Royal S. Copeland submitted "A National Plan for American Forestry," which recommended a large extension of public ownership of forest lands combined with more intensive management.[82] The Copeland Report combined with other efforts to spur intensive study of forest management during the 1930s. While the results were mixed, they fueled the further connection of forest management to other issues, including soils and wildlife. FDR withdrew public lands for a variety of purposes, including the conservation of energy and wildlife resources. In 1936, FDR called a wildlife conference intended to develop a North American program for the advancement of wildlife restoration and conservation.[83] As part of the National Works Program, the project in wildlife conservation sought to maintain adequate seed stock of wild animals and, when possible, to preserve and protect environments that aided the species' preservation.

Regarding water resources, FDR emphasized the control of the floods that had plagued many watersheds in the early 1900s. Of course, this effort to actively conserve and manage water resources lay at the foundation of Tennessee Valley Authority (TVA), one of the New Deal's large-scale conservation projects. The tools of such management severely altered the Tennessee River watershed through the use of dams. However, this river control was part of a multiuse watershed development plan that included soil and forest conservation, which also helped to subdue runoff and flooding while restoring soils for agriculture. In addition, New Deal projects emphasized fish and wildlife restoration associated with watersheds as well as recreational development. By the end of the New Deal decade, Harold Ickes, Secretary of the Interior, stated simply, "there can be no question about the advance we have made upon the conservation front."[84] Historian R.E. Owens reports that throughout his presidency, FDR's conservation policy "never lost favorable publicity."[85] Throughout the New Deal, these policies sought to define and take out of use those lands that were marginal for agricultural production. They clarified the social consequences of poor environmental practices and began to suggest ways of improvement.

How did the president remain connected with such policies? A variety of methods were used to provide FDR ownership of the New Deal's conservation policies. For instance, at moments such as dedications, FDR demonstrated his ability to allow modern technology to circumvent his personal disability. A hallmark image of FDR's presidency became the president seated in the White House while being linked to various grand openings or other extravaganzas by a telephone or an electronic lever or button. Largely ceremonial, such buttons might allow the president to turn on electronic devices to mark the opening of carnivals, dams, or new facilities. Such acts were about control and FDR's assertion of ownership over specific accomplishments; however, they must also be appreciated for their ability to circumvent the man's personal immobility. Such uses of technology, including radio for "fireside chats," allowed the most immobile president in American history to also be one of the most ubiquitous. In this fashion as well as in the very accomplishments of TVA, FDR transcends his physical limitations to implement a new environmental ethic for the American people.

THE VIEW FROM SPRINGWOOD

While Franklin greatly admired the environmental heroes of the previous generation, he presided over an era of applied environmentalism. He performed in public service during the vanguard effort to apply federal authority to environmental regulation and planning. While many New Deal agencies involved putting Americans to work by managing or harvesting natural resources, the policies also grew from a modern perspective based in the science of ecology: simply, that economic problems are indications of other social, ecological, or cultural difficulties and that such areas of concern must also be involved in the solutions to such difficulties. The ethic behind such ideas is based around reducing waste; however, an important new emphasis explored various dynamics shaping wasteful habits or trends. While many other thinkers also helped to define New Deal conservation policies, the majority of these initiatives fell directly in line with the ethics and values that Franklin exhibited earlier in his life. The New Deal conservation agencies are the best indications of this change. However, many observers believed that the basic concepts of ecology had forever altered land planning. For instance, in 1939 *Harper's* published, "Science and the New Landscape" by Paul B. Sears, the revolutionary ecologist who wrote *Deserts on the March* in 1935. He explains to readers that landscapes are no longer inert or changeless objects. The sciences of ecology and anthropology, which were used at some level in most New Deal land planning, "have taught that man himself is not a watcher, but like other living things, is a part of the landscape in which he abides. This landscape, including its living constituents, is an integrated whole."[86] He concludes by saying that scientists of the future must continue to work with an awareness of social and economic processes—but there is another hurdle: "The scientist must be aware of the relation of his task to the field of aesthetics."[87] In the opinion of the New Deal, the president must then be aware of the viewpoints of the scientists. A new aesthetic of efficiency had taken shape with the modern landscape serving as its product and symbol. More than any other individual, FDR bore responsibility for orchestrating the application of this intellectual shift—a new environmental ethic—on to the American landscape.

When FDR died in office in 1945, he had left careful plans for the location of his body, his papers, and his legacy. With its sumptuous view of the Hudson, Springwood became the final resting spot of the complex environmentalist. One figure worked tirelessly to prepare the resting spot. The gardener, Plog, planted and maintained the rose garden that remains famous today. Of more importance to Franklin, though, was the view from this lovely garden. Once, it had been the Hudson that overwhelmed visitors to the estate. As the twentieth century proceeded, a more significant draw may have been the man-made shield that has come to block the house's view of the river: Roosevelt Woods, as it is now called, complete with its spectacular stands of tulip poplar. The forest stands as the best monument to the man and the ethic that drove many of his actions on the land.

NOTES

1. Franklin Delano Roosevelt, "First Inaugural Address," Washington DC, March 4, 1933. The author wishes to thank Jennifer Delton, Michael Birkner, Clyde Black, Neil Maher, Sara Pritchard, Paul Sutter, John Opie, and Donald Worster. Much of this research was completed as a Beeke-Levy Research Fellow at the FDR Archive and Library and National Park Service Archives, Hyde Park, NY. In addition to the foundation, debts were incurred from the staff of that facility, Skidmore College and Penn State University. Other research was made possible by the assistance of the staff of the National Archives, particularly Arlene Royer at the Southeastern Branch of the National Archives in College Park, GA . Finally, the work of Charles W. Snell, a former archivist with the National Park Service, bears special mention. Without his keen eye of documents and careful recording, it is likely the Log books cited in this article would not have been rediscovered for a considerable time.

2. In this article, FDR is used to discuss Roosevelt during his presidency. In an effort to distinguish between the politician and the man, the narrative refers to Roosevelt as Franklin during his younger years.

3. The most effective discussion of FDR's life with polio is Richard Thayer Goldberg, *The Making of Franklin D. Roosevelt* (Cambridge: Abt Books, 1981).

4. There is no shortage of biographies of Franklin during each stage in his life. For this analysis, I have used Geoffrey C. Ward, *A First Class Temperament* (New York: Harper and Row, 1989). Ward's volume specifically concerning Franklin's very early life was consulted but not cited for this article. Geoffrey C. Ward, *Before the Trumpet* (New York: Harper and Row, 1985).

5. This sensibility was well known in Jefferson's writings. Most famously, he referred to agriculturalists as "the chosen people of God." For a discussion of these points, see Charles A. Miller, *Jefferson and Nature* (Baltimore: Johns Hopkins University Press, 1993).

6. Theodore was second cousin to Franklin and uncle to Eleanor. He gave Eleanor away at her wedding to Franklin. See Paul Russell Cutright, *Theodore Roosevelt, the Making of a Conservationist* (Chicago: University of Illinois Press, 1985).

7. *Nature's Nation* (Cambridge: BelkNap Press of Harvard University Press, 1974).

8. For instance, see David Lowenthal, *George Perkins Marsh: Prophet of Conservation* (Seattle: University of Washington Press, 2000).

9. In reality, of course, upper- and middle-class women were a significant constituency behind early environmental efforts. However, it was much easier for Americans to conceive of the worth of the outdoors if it were being used for a bona fide recreational purpose. These purposes were most often the hunting, fishing, and camping activities that were most closely linked to males.

10. Stephen Fox, *The American Conservation Movement* (Madison: University of Wisconsin Press, 1981).

11. Roderick Nash, *Wilderness and the American Mind* (New Haven: Yale University Press, 1982), 141.

12. Samuel P. Hays, *Conservation and the Gospel of Efficiency* (Pittsburgh: University of Pittsburgh Press, 1999), 264.

13. Richard W. Judd, *Common Lands, Common People* (Cambridge: Harvard University Press, 1997).

14. See Nash, *Wilderness* or Peter J. Schmitt, *Back to Nature* (Baltimore: Johns Hopkins University Press, 1990).

15. Richard White, essay "Are You an Environmentalist or Do You Work for a Living?: Work and Nature," contained in William Cronon, *Uncommon Ground* (New York: Norton, 1996), 171–73.

16. Many historians of this era have followed a paradigm that divides environmental activities into preservation, which is largely aesthetically motivated, and conservation, which is extremely utilitarian. See, e.g., Nash, *Wilderness* and Fox, *American Conservation Movement.*

17. Historians have explored this connection. See, e.g., A.L. Riesch Owen, *Conservation Under F.D.R.* (New York: Praeger, 1983). Also, his speeches and writings have been collected by Edgar B. Nixon, *Franklin D. Roosevelt & Conservation, 1911–1945* (Washington DC: National Archives and Records Service, 1957). While each source is extremely useful, no single text brings the personal interests of FDR together with the New Deal's watershed conservation policies.

18. For a definition of environmental ethics, see Aldo Leopold, *A Sand County Almanac* and Zimmerman, Michael E., ed., *Environmental Philosophy.*

19. My suggestion here is that due to the difficulty of his immobility, the desire that drove FDR's outdoor activity must be compounded in significance. The organization and exertion required indicates that the desire must have been particularly compelling and the individual's resolve exceptional.

20. The most compelling description of Franklin's upbringing at the Hyde Park estate may be Ward's chapter "The Place," *A First Class Temperament,* 263.

21. For a general description of life along the Hudson, see Carl Carmer, *The Hudson* and Thomas S. Wermuth, *Rip Van Winkle's Neighbors.*

22. Ward, *A First Class Temperament,* 500.

23. *Franklin D. Roosevelt and Hyde Park: Personal Recollections of Eleanor Roosevelt,* 8.

24. Ibid., 8.

25. These entries combine with William A. Plog's "Memorandum Book" to offer insight into Franklin's use of the land during the first decade of the 1900s. One helpful guide is the catalogue created by Charles W. Snell, "Franklin D. Roosevelt and Forestry at Hyde Park, New York, 1911–1922," collected May 20, 1955, which is contained at the Presidential Library.

26. Snell, "Franklin D. Roosevelt and Forestry at Hyde Park," iii.

27. "The Archibald Rogers Crumwald Hall" Papers, Roosevelt Presidential Library.

28. Snell, "Franklin D. Roosevelt and Forestry at Hyde Park," 1–3.

29. Roosevelt, Elliott, ed., *FDR, His Personal Letters, Early Years* (New York: Duell, Sloan and Pearce, 1947), 459.

30. *Franklin D. Roosevelt and Hyde Park: Personal Recollections of Eleanor Roosevelt,* 12.

31. FDR to Hendrek William van Loon, February 2, 1937, President's Personal File.

32. Lost in the Roosevelt archives, the leather-bound ledger was initially located by Snell in the 1950s and then lost again. The "Farm Journal," as the ledger is now catalogued lacks the philosophical treatise for which many FDR researchers may have hoped; however, it clearly shows the young Roosevelt to be deeply interested in the management of the natural and economic life of the Springwood Estate.

33. Ward, *A First Class Temperament,* 376–77.

34. In the fall of 1907 Franklin joined Carter, Ledyard and Milburn, one of New York's most prestigious law firms. Many scholars credit his work in the Municipal Court of New York with increasing his interaction with and sensitivity to ordinary citizens. In 1910, he ran as a Democrat for state senator in a traditionally Republican district. In the New York State Assembly Franklin perfected the political skills that made him one of the smoothest handlers of people in American history. As a young, independent, progressive Democrat Franklin supported Woodrow Wilson in 1912. In March 1913, Franklin was sworn in as Wilson's assistant secretary of the Navy. Franklin favored the nation's early involvement in World War I, and therefore he grew frustrated with Wilson by the end of the decade. In 1920, he ran for vice president on the ticket of James Cox.

35. At this juncture, FDR also was named vice president in charge of the New York office of Fidelity and Deposit Insurance Company. Recognition of his name had grown significant enough that he could attract such a largely ceremonial post.

36. Ward, *A First Class Temperament,* 559.

37. Ward and others have seized on this description of young Franklin. Inferred in its usage, of course, is a level of familial stature; however, in addition the phrase seems to describe well the personal characteristic that made FDR excel as a leader under pressure.

38. Ward, *A First Class Temperament,* 580–81.

39. Ibid., 597. On August 10, Franklin took his family sailing. From the boat, they spotted a forest fire on shore. They stopped, fought the fire successfully, and then returned to the house for a swim in a nearby lake and later in the ocean. Franklin returned to the cottage, felt a chill, and went to bed. The following morning, Franklin awoke to discover his left leg dragging. It had become too weak to sustain his weight. His fever rose to 102 degrees by evening and the pain in his legs and back grew unbearable. It would later be discovered that Franklin had contracted polio and he would never regain the use of his lower extremities. In fact, after initially losing their use, he regained movement to his arms only through vigorous rehabilitation.

40. This historical source is unfinished and ripe with the writer's own educated guesses (and a lack of citations). Most often, the guides to his story were

trees: "The old oak tree in front of the Library and on the lot south of the Avenue must, of course, have grown up under field conditions and this existed only where Indians had cleared the land and cultivated it." Hyde Park folder, FDR Presidential Library, with cover letter from Grace Tully, May 15, 1945.

41. Ward, *A First Class Temperament,* 169–70.
42. Correspondence, FDR to Plog, July 20, 1937, Franklin Roosevelt and ESF— Training a Forester–President, Alumni Newsletter, SUNY College of Environmental Science and Forestry, vol. 83 (Winter 1984).
43. Snell, p. 2. This material is also discussed in Thomas Patton and Neil Maher, unpublished Ph.D dissertation New York University, 2000.
44. Snell, Franklin D. Roosevelt and Forestry at Hyde Park, FDR Library.
45. Cox, *Well-Wooded Land,* 215.
46. Brown, *New York Times,* April 10, 1933. Also see Brown's article "The President Practices Forestry," *Journal of Forestry,* vol. 41, No. 2 (February 1943).
47. These records appear in many sources. Snell is likely the source for it.
48. Letter to George Pratt, November 25, 1922. Roosevelt Presidential Archive.
49. The extent of this correspondence is fascinating. FDR to Gentlemen at Conservation Commission (CC) Albany FDR, April 5, 1924: "I had planned to put these trees into my own transplanting bed for one year or two years before distributing them to the 4 or 5 people in the Grange who wanted to try them out as I think they will take ore interest in the trees if they look a little bigger when they are put in and will thereby be induced to keep them free from brush." Response to C.R. Pettis (Supt. of CC of NY), April 22, 1924: "I am doing everything possible to interest the local granges in planting and I have found that the best approach to the average farmer is through persuading him that a properly cared for wood lot, together with young pine plantings will enhance the selling value of his farm far more than the cost of trees." Gardener will plant trees when they arrive. He hopes Pettis will come by to his "planting."
    April 24, 1924 from Pettis: "I note what you say in regard to interesting the local granges, and it is our idea that the proper handling of the farm wood-lot is of as much importance as is the reforesting of the idle lands." Pettis then asks FDR to help in setting up a county demonstration area.
50. Various correspondence, 1922–1928, collected at Roosevelt Presidential Library.
51. Correspondence with Conservation Commission, 1928. Roosevelt Presidential Library.
52. FDR Correspondence, April 5, 1924. Roosevelt Presidential Library.
53. Correspondence from Pettis, April 24, 1924. Roosevelt Presidential Library.
54. Correspondence from Newton Armstrong, August 30, 1928. Roosevelt Presidential Library.
55. Personal correspondence from this period reflects these matters. See also, Ward.
56. The resort's purchase price, Eleanor wrote, would threaten their ability to pay for their children's education and other necessary expenses. Undaunted, Franklin committed to the $195,000 purchase price.

57. See Snell.
58. Correspondence Brown to Roosevelt, February 4, 1930. Roosevelt Presidential Library.
59. Irving Isenberg, "Management Plan for Kromelbooge Woods at Hyde Park, NY for the Period, 1931–1941," FDR Library, Hyde Park, N.Y. 4.
60. "My Day." October 9, 1946, Eleanor Roosevelt Papers, Roosevelt Library.
61. For readers wishing to explore Roosevelt's efforts in forestry more closely, see Thomas W. Patton, "Forestry and Politics: FDR as Governor of New York," *New York History* (October 1994): 397–418.
62. See Patton.
63. John Morton Blum, *Roosevelt and Morgenthau* (Boston: Houghton Mifflin, 1970), 10–20; 24.
64. Patton, 406.
65. "Our experience is that we cannot depend upon the owner of land to do a large amount of reforestation work. If he owns much of this land, he is generally too poor to reforest it. If a land utilization study indicates that there are large areas which ought to be reforested, the job must probably be done by the county or state." Patton, 407.
66. New York Times, July 7, 1932, p. 13.
67. Patton, *Journal of Forestry* (February 1994): 17.
68. Correspondence, April 8, 1942. Roosevelt Presidential Library.
69. For a discussion of modernity see Marshal Berman, *All that is Solid Melts into Air*, David Harvey, *The Condition of Postmodernity*, and Steven Kern, *The Culture of Time and Space*.
70. See Maher.
71. In the paradigm argued by Leo Marx in *The Machine in the Garden*, American culture has largely been constructed through an ongoing struggle between the pastoral and technology, or civilization. American generations have struggled to reconcile this situation by constructing a "middle ground" between the two. Such a subjective construct, of course, varies with cultural eras.
72. Hays, *Conservation and the Gospel of Efficiency*, 265.
73. In *The Machine in the Garden*, Leo Marx introduced a paradigm of American development in relationship to technology and undeveloped nature. Between the poles of the machine and the garden, he believed there was a "middle ground" of reconciliation that Americans at times achieved.
74. Donald Worster, *Nature's Economy* (New York: Cambridge University Press, 1977), 233.
75. Ibid., 219–20.
76. Ibid., 232.
77. Worster, 232.
78. Hays, *Conservation and the Gospel of Efficiency*, 271.
79. Ibid., 275, "first conservation movement experiments with application of the new technology to resource management. Requiring centralized and coordinated decisions, however, this procedure conflicted with American political institutions which drew their vitality from filling local needs."
80. Owen, *Conservation Under F.D.R.*

81. Ibid.
82. Ibid.
83. Ibid.
84. Ibid.
85. Ibid.
86. Paul B. Sears, "Science and the New Landscape," *Harper's* (July 1939): 207.
87. Ibid., 216.

CHAPTER 3

"A Conflux of Desire and Need": Trees, Boy Scouts, and the Roots of Franklin Roosevelt's Civilian Conservation Corps

Neil M. Maher

On April 7, 1933, just two weeks after President Franklin Roosevelt signed the Civilian Conservation Corps (CCC) into law, Henry Rich of Alexandria, Virginia became the first American citizen to enroll in the New Deal program. After spending ten days at Fort Washington, an army conditioning camp just outside the nation's capital, Rich and approximately two-hundred other young men boarded buses and traveled south through Luray, Virginia into nearby George Washington National Forest. After hiking through the forest up to a valley in the Massanutten mountains, the enrollees arrived at a site selected by the United States Forest Service. As dusk settled and the weather grew cold, the men ate a hot meal prepared from a make-shift kitchen and stretched out on the ground with blankets for the night. In the morning, before they began clearing the area of undergrowth and constructing their new living quarters, the young men took a vote and decided to name their camp, the first in the nation, after the president who created the Corps. By week's end these CCC enrollees had constructed an elaborate wooden sign, ten feet tall by ten feet wide, that announced "Camp Roosevelt" to all visitors.[1]

Although the nation's first CCC enrollees named their camp in honor of Franklin Roosevelt, throughout the 1930s dozens of individuals laid claim to having conceptualized what was often heralded as the New Deal's most popular program. British forester Richard St. Barbe Baker was perhaps the first of such claimants, arguing that he suggested the CCC idea to Franklin Roosevelt during a meeting in Albany just before the presidential election of 1932. Four years later in the fall of 1936, an unemployed electrician and father of two named Joseph Wilson also laid claim to the Corps idea, and drove from his home in Atlanta, Georgia to Washington, DC where he forced his way into the White House and demanded, unsuccessfully, to take the matter up with the president. Finally, a Brooklyn, New York resident named Major Julius Hochfelder wrote several letters to CCC headquarters stating that he was in fact the ideological creator of the Corps, and should therefore receive a job with the New Deal program. Such claims were nothing new to CCC director Robert Fechner. "We have letters from a large number of individuals . . . who feel they first conceived the idea of this organization," he wrote to one of the Hochfelder's supporters. "I merely point out these things so that you may know that Major Hochfelder is not alone in thinking that he was entitled to some reward for suggesting the CCC plan to the president."[2]

Similar to the debate between these individuals, contemporaneous accounts of the Corps' genesis also differed. Articles and books written during the 1930s generally highlighted four influences behind the CCC's birth. One of the most common explanations concerned the essay titled "The Moral Equivalent of War," written by Harvard philosopher William James in 1906. According to James, "instead of military conscription" the United States should adopt "a conscription of the whole youthful population to form for a certain number of years a part of the army enlisted against Nature."[3] Other accounts of the Corps written during the Great Depression emphasized as well the establishment after World War I of youth work programs in several European countries including Bulgaria (1921), Switzerland (1924), and especially Germany (1925), whose German Labor Service was most often compared to Roosevelt's CCC.[4] Rising juvenile delinquency during the early years of the Great Depression was yet another cause often cited by newspaper reporters for the creation of the Corps. An article in *The New Republic,* for instance, stated that the president established the CCC "to prevent the nation's male youth from becoming semi-criminal hitch-hikers."[5] And finally, commentators during the 1930s often portrayed Franklin Roosevelt's practice of conserving natural resources prior to becoming president as a major factor in his decision to establish the CCC.[6]

During his presidency, Roosevelt did little to clarify this debate concerning the origin of the Corps. Not only did he claim that he had never read William James's essay, but he also failed to acknowledge the influence of other nations' youth work programs on his own thinking. Instead, Roosevelt repeatedly responded to queries concerning the ideological roots of the Corps by referencing similar programs he initiated while governor of New York that likewise put unemployed men to work in state parks and forests, thus pushing the origin question back in time rather than answering it.[7] When forced to respond, as when *Time* magazine editor I. Van Meter personally wrote the White House in 1939 for a cover story on the CCC, Roosevelt was equally evasive. The president "cannot find that the idea of the Civilian Conservation Corps was taken from any one source," wrote Roosevelt's private secretary, Marguerite LeHand, to *Time* magazine. "It was rather the obvious conflux of the desire for conservation and the need for finding useful work for unemployed young men."[8]

This conflux of desire and need resulting in the creation of the CCC is central to understanding the New Deal's impact on the American conservation movement.[9] Not only does the Corps idea illustrate the president's unique brand of conservationist thinking during his first term in office. Perhaps, more importantly, it also suggests the New Deal ideologies that helped transform conservation as a whole during the Great Depression in ways that foreshadowed the post–World War II environmental movement. To fully grasp Franklin Roosevelt's environmental legacy, then, we must chart the intellectual geography of the Corps' creation.[10]

In mapping the ideological landscape that gave rise to the CCC it is best to begin with the "congressional action" that created the New Deal program in March of 1933. Although Congress passed the bill establishing the CCC on March 31, 1933, the events creating it began several weeks earlier when on March 9 Franklin Roosevelt sketched out rough plans for putting 500,000 unemployed men to work on conservation projects throughout the country. Over the next few days the president fine-tuned his thinking, deciding to limit enrollment in the program to young men between the ages of 18 and 25 who were willing to send 25 of their 30 dollar monthly pay home to their families, all of which had to be listed on state relief registers. Roosevelt also decided to house these "enrollees," as they were called, in 200 man camps located in national and state parks and national forests, and to run the New Deal program cooperatively. While the Department of Labor would coordinate enrollee recruitment and the Department of War would be responsible for the daily functioning of the CCC camps, the Department of Agriculture would supervise conservation projects in national forests while Interior oversaw work performed in national and state parks. After outlining these ideas to brain truster Raymond Moley on March 14, the president asked the secretaries of War, Interior, Agriculture, and Labor to coordinate plans for putting the proposed program into operation and to report back to him. One week later on March 21, Roosevelt formally asked Congress to establish the CCC.[11]

Because it represents his most developed thinking regarding the Corps prior to its creation, Roosevelt's March 21 congressional address serves as a useful guide through the murky intellectual terrain surrounding the program's origins. Titled "Relief of Unemployment," the president's address began by drawing attention to the most obvious crisis then gripping the nation: the fact that thirteen million, or one in four working-age citizens, remained jobless. "It is essential to our recovery program," wrote Roosevelt, "that measures be immediately enacted aimed at unemployment relief." After warning members of Congress that the "enforced idleness" associated with joblessness threatened the "spiritual and moral stability" of the nation, and reminding them that the "overwhelming majority of unemployed Americans . . . would infinitely prefer to work," the president proposed three types of work-relief initiatives, one of which was the CCC. "I estimate," he wrote to Congress regarding the proposed Corps, "that 250,000 men can be given temporary employment by early summer if you give me authority to proceed within the next two weeks."[12]

Although the congressional address suggested that the unemployment emergency was paramount in Roosevelt's thinking, it was not the only crisis on his mind early in 1933. The state of nature also alarmed the president, and he expressed this as well in his March 21 message. After warning of the dangers posed by joblessness, Roosevelt's address directed Congress's attention to "the news we are receiving today of vast damage caused by floods on the Ohio and other rivers," due in large part to deforestation along their banks. The president dismissed the notion that these disasters were natural and instead blamed human negligence, arguing that the floods occurred because "national and state domains have been largely forgotten in the past few years of industrial development." To make up for such neglect, the federal government had to take action to "conserve our precious natural resources" located on these important public lands. The CCC was Roosevelt's first step in this process. "I propose to create a civilian conservation corps," he wrote to Congress, "to be used in simple work, not interfering with normal employment and confining itself to forestry, the prevention of soil erosion, flood control and similar projects."[13]

As expressed in his message to Congress, in creating the CCC the president responded to two national crises, one involving unemployed youths and the other relating to a degraded natural environment. Yet although his congressional address helps identify these two strands of concerns, it does little to answer the more important question regarding how these ideological tributaries flowed together in Roosevelt's early New Deal thinking. This is especially important considering that earlier in the century such concerns had been quite distinct. During the Progressive era, for instance, the conservation movement promoted the efficient use of natural resources such as timber, soil, and water, but rarely, if ever, concerned itself with how such increased efficiency might affect workers' lives.[14] Similarly, progressives interested in unemployment reform refrained from promoting conservation as a means of decreasing idleness in youth.[15] How, then, did Roosevelt decide to fuse the needs of nature with those of young jobless men in March of 1933? Part of the answer lies in the evolution of Franklin Roosevelt's ideology regarding both conservation and youth relief prior to becoming president, a period when these two strands of thought were anything but in conflux.[16]

Franklin Roosevelt first experienced the deterioration of natural resources at a young age while growing up on his family's estate in Hyde Park, New York. Located approximately half-way between New York City and Albany, the 1,200-acre parcel of land sloped steeply upward from the eastern bank of the Hudson River to a level bluff on top of which sat the family home. Although the Roosevelt family acquired the property, which they called Springwood, in the early nineteenth century, Dutch settlers had worked much of the estate's land for over 200 years. This became all too obvious to Roosevelt when in 1910, at the age of 28, he took over management of the property from his mother, Sara. After learning from estate records that his ancestors had grown prize-winning corn at Springwood in 1840, Roosevelt became understandably concerned that 70 years later the property produced only half of what it had in the mid-nineteenth century. He also became alarmed at the large gullies that had formed and continued to widen down along the property's steepest slopes, slopes that had been cleared decades before for cultivation. With every rain these gullies washed fertile topsoil off the property down into the Hudson River.[17] The rest of the estate suffered similarly; yields of grain, pasturage, fruit, and vegetables were far below average because the rocky Hudson Valley soil had long since passed its peak. As Roosevelt put it, "I can lime it, cross-plough it, manure it and treat it with every art known to science, but it has just plain run out."[18]

Similar to his encounters with a deteriorating natural environment, Roosevelt's experiences conserving natural resources also began early in life. In 1891 while vacationing with his family in Europe, Roosevelt bicycled through the countryside near Bad Hauheim, Germany, and encountered a small town with a large municipal forest on its outskirts. From discussions with local residents, the nine-year-old cyclist learned that the forested tract had been carefully managed for the last 200 years to yield an annual timber crop that offset the town's expenses. "The interesting thing to me, as a boy even, was that the people in the town didn't have to pay taxes," remembered Roosevelt many years later. "They were supported by their own forest."[19] It was this type of thinking that Roosevelt brought back with him to Hyde Park.

Roosevelt began conservation efforts at Springwood when he took over the day-to-day operation of the estate in 1910. He first bought several adjacent farms, which had also been depleted by generations of poor husbandry, and in 1911 asked foresters at the University of Syracuse's State College of Forestry to develop a reforestation program for the enlarged estate "in the hope that my grandchildren will be able to raise corn again—just one century from now."[20] On the foresters' recommendations, Roosevelt planted a few thousand trees in 1912 and continued an annual planting regimen until his death in 1945, at which time he had supervised the planting of more than half a million trees covering 556 acres. In 1933 alone, the year Roosevelt created the CCC, he planted 36,000 trees at Hyde Park. Thirty-two species were included in Roosevelt's plantings, the most common being Norway and Canadian spruce, Scotch, Norway, and white pine, and Tulip poplar, which was his favorite tree. He introduced a number of exotics as well, such as European and Japanese larch, Sitka spruce, Douglas fir, and Western yellow pine.[21] In light of this extensive effort, it is not surprising that each year when voting in Hyde Park Roosevelt listed his occupation as "tree grower."[22]

Although he referred to reforested plots similar to those at Hyde Park as "the most potent factor in maintaining nature's delicate balance," Roosevelt planted trees on his property for economic, not ecological, reasons. According to Nelson Brown, the Syracuse University forester who oversaw the plantings at Springwood, reforestation for Roosevelt was "not just a passing fancy or plaything—it is a very realistic and practical business endeavor" that often turned a profit. Throughout the 1920s and early 1930s, Roosevelt did indeed make small returns by selling fuelwood and sawlogs locally, as well as crossties cut from his woodlots to the New York Central Railroad, which ran a train line along a narrow strip of land between the Hudson River and the Roosevelt property. Roosevelt's most publicized foray into commercial forestry, however, began in 1926 when he planted the first of many plots of Norway spruce, which at the time were used widely as Christmas trees. Only nine years later he cut and sold 132 of these trees for a modest profit of 134 dollars and 55 cents, and in 1937 harvested 1,000 trees for a total of 480 dollars. All told, the Johnny Appleseed of Springwood sold several thousand Christmas trees worth almost as many dollars during the 1930s and 1940s.[23] During the early years of the latter decade Roosevelt also harvested more than 2,000 mature hardwood trees that yielded 445,000 board feet of timber for a net gain of more than 60 dollars per acre.[24] "He realizes," explained Nelson Brown of Franklin Roosevelt, "that aesthetic considerations are in order about his home, but out in the forest, they play a very small—in fact, a negligible part." According to Brown, Roosevelt continually inquired about the price of various forest products and weighed the relative merits of growing species for pulpwood, sawlogs, fuelwood, crossties, and Christmas trees. "He believes in taking advantage of favorable market conditions when available," added the Syracuse forester.[25]

Just as he transported what he learned in Bad Hauheim, Germany, back to Hyde Park, Roosevelt brought the conservation knowledge he gained at Springwood with him to the New York State Senate when he began serving in that body in 1911. Because of his statewide reputation as a "tree grower," Roosevelt's first appointment was as chairman of the Senate's Forest, Fish, and Game Committee. In that capacity he publicized a number of threats to the state's natural resources and introduced eight bills aimed at conserving them, including legislative initiatives regulating fishing, hunting, and the development of water power.[26] By far Roosevelt's most intense senatorial battle, however, involved a bill he proposed in January of 1912 that would, among other things, allow the state to regulate timber harvests on private land. The Roosevelt–Jones Bill, as it was popularly called, arose in response to a disturbing trend then occurring in upstate New York's Adirondack Forest Preserve. In 1895 when a state constitutional amendment created the preserve, the legislature in Albany delineated its geographical boundaries by "drawing" a "blue line" on maps around twelve counties in the Adirondack mountain region of upstate New York. The result was a 3.3 million-acre reserve comprising both state and privately owned land in nearly equal parts. Private individuals, associations such as the Adirondack Mountain Club, and numerous lumber companies, all owned large parcels of land within the boundaries of the state park. More alarming to Roosevelt was that although the constitutional amendment declared state-owned property within the preserve to be "forever kept as wild forest land," the law did not apply to the park's privately owned parcels, which during the first decade of the twentieth century became increasingly deforested.[27] The Roosevelt–Jones Bill was aimed at regulating, not outlawing, this cutting on private lands in the Adirondacks and throughout New York.

Roosevelt defended his bill in the Senate and across the state in the same manner he defended his conservation efforts at Hyde Park; he argued that it made good economic sense. "It is an extraordinary thing to me," he wrote to one constituent concerning the lumbermen opposing his initiative, "that people who are financially interested should not be able to see more than about six inches in front of their noses."[28] By regulating logging on private property, Roosevelt argued, the Roosevelt–Jones Bill would reduce water runoff and soil erosion on adjacent state-owned land and thus assure the long-term financial security of the state's forests. To help promote the bill, Roosevelt invited the well-known chief of the U.S. Forest Service Gifford Pinchot to lecture before the assembly in Albany. Pinchot illustrated his talk with two lantern slides of a valley in China: the first slide was of a painting from the year 1500 depicting a lush landscape covered with trees, crops, and numerous signs of human habitation while the second, a photograph taken four centuries later, portrayed the same landscape void of both vegetation and humanity. The message was all too clear; poor land use was suicide. The presentation not only confirmed Roosevelt's belief in the economic necessity of conservation but made such an impression that he publicly referred to Pinchot's slides numerous times throughout the rest of his political career.[29]

Although the senator from Hyde Park justified the Roosevelt–Jones Bill on economic grounds, much as he defended his tree-planting practices at Springwood, during this political debate Roosevelt also began formulating a social component to his conservationist ideology. He first expressed this in March of 1912 while delivering a speech in support of the bill before the Troy, New York, People's Forum. Pointing to the country's founding fathers, Roosevelt told the crowd that the basic thrust of modern history had been the struggle and attainment of individual liberty. In the same breath, however, he warned that this "individual freedom was inevitably bound to bring up many questions that mere individual liberty cannot solve." One such question concerned the most efficient use of natural resources. To make his point Roosevelt described the lantern slides Gifford Pinchot had shown in Albany just weeks before, and told his audience in no uncertain terms that "this is what will happen in this very State if the individuals are allowed to do as they please with the natural resources to line their own pockets." To avoid such a fate, Roosevelt proposed what he called a new social theory that posited community cooperation as more economically efficient than individualism. "To put it in the simplest and fewest words," he explained, "I have called this new theory the struggle for liberty of the community rather than liberty of the individual." Communities, he argued, must have the freedom to protect their interests from short-sighted individuals such as Adirondack lumbermen. The Roosevelt–Jones Bill gave the "community" of New York residents this freedom, and only through its passage would the development and use of the state's natural resources "be put on the most economical and at the same time the most productive basis."[30]

Even though upstate lumber interests disagreed with Roosevelt and successfully convinced the Albany legislature to strike that section of the 1912 bill permitting the state to regulate logging on private land, other individuals during the Progressive era were more sympathetic. As industrialization's devastating impact on the nation's water, soil, and timber supplies became increasingly apparent during the late nineteenth and early twentieth centuries, reformers took concerted action to manage resources more scientifically for future use. This desire to produce natural resources rationally—what historian Samuel Hays has called "the gospel of efficiency"—was in fact the central tenet of the progressive conservation movement.[31] The Newlands Reclamation Act of 1902, the creation of the U.S. Forest Service in 1905, and the passage of the Weeks Act in 1911, which provided for federal acquisition of private forest land, were only the most well-known attempts by conservationists to improve resource production. To these progressive reformers, nature was a collection of resources waiting to be efficiently developed.[32]

Unlike other reform efforts during the Progressive era, an emerging middle class in search of order did not orchestrate the conservation movement. Instead, a triumvirate of scientific professionals, government bureaucrats, and businessmen involved in resource extraction directed the reform effort from above.[33] Perhaps no single individual better epitomized the movement, or its elite composition, than Gifford Pinchot. Born in 1865 to a wealthy Connecticut family, Pinchot grew up in Paris and Pennsylvania, graduated from Yale in 1889, and then traveled to France and Germany to study forestry. Upon his return to the United States in 1892 he began promoting the scientific management of the country's timber supplies, and practiced what he preached, first as timber manager at Biltmore, the Vanderbilt's family estate near Asheville, North Carolina, and later as a forester for the federal government. Throughout his career, Pinchot viewed nature in strict utilitarian terms. For instance, in his autobiography he defined conservation as "the development and use of the earth and all its resources for the enduring good of men."[34] To determine how best to produce and use these resources, Pinchot also developed a simple formula for the Forest Service to follow. In a 1905 memo written soon after becoming chief forester, Pinchot explained that when faced with conflicting interests regarding the production of a natural resource, such as the situation Roosevelt encountered in the Adirondacks, the question should "always be decided from the standpoint of the greatest good of the greatest number in the long run."[35] This principle not only became the guiding philosophy of the Forest Service but of the great majority of progressive conservationists as well.[36]

While Pinchot served as the unofficial spokesperson for conservationists, the movement itself was suffering from growing pains during the Progressive era. Rumblings began surfacing during the late nineteenth century when another point of view concerning the natural environment emerged and gained adherents. The most well-known promoter of this alternative vision was John Muir, foremost lobbyist for the creation of Yosemite National Park in 1890, founder of the Sierra Club in 1892, and an indomitable outdoorsman whose nature writing propelled him to minor literary celebrity around the turn of the century. Muir's early life could not have been more different than Gifford Pinchot's. Born in Dunbar, Scotland in 1838 to overbearing Calvinist parents, Muir moved with his family at the age of 11 to a homestead on the central Wisconsin frontier. Unlike Pinchot, who studied at the best schools in America and abroad, Muir was a self-taught naturalist who attended the University of Wisconsin for only two years. Before leaving Madison, however, he encountered the writings of Wordsworth, Emerson, Thoreau, and a lesser-known minister named Walter Rollins Brooks. From that moment on

Transcendentalism influenced the Scotsman's philosophy regarding the
natural world. According to Muir, natural objects were "the terrestrial manifestations of God," and nature itself was "a window opening into heaven, a
mirror reflecting the Creator." Leaves, rocks, and bodies of water were "sparks
of the Divine Soul."[37] Wary of Gifford Pinchot's utilitarian view, Muir
believed in preserving nature's beauty for its own sake as well as for the
spiritual sake of humankind. And although Muir rarely used the term "conservationist" when describing himself, during the Progressive era his preservationist philosophy and Pinchot's utilitarian vision existed side-by-side in an
uneasy alliance. While Pinchot embodied the mainstream, moderate, professional wing of conservation, Muir and his followers represented the radical
amateur who served as the movement's conscience.[38]

Muir and Pinchot first met in 1893 and became friends three years later
while working and camping together as part of a federally sponsored survey
of western national forests, which at the time were called "reserves." Members
of the expedition were responsible not only for examining the reserves but
also for recommending a federal policy for their management. During the
trip the two men found much in common, and often left the camaraderie of
the evening campfire to discuss their mutual love of the outdoors and the
future of the nation's forests, which both agreed needed protection from private, unscientific development. Each morning, Pinchot recalled years later,
"we sneaked back like guilty schoolboys" to camp to rejoin the other members of the expedition.[39] Yet when the survey members began preparing their
report, the limits of the two men's common interest became apparent.
Whereas Muir recommended that the federal government preserve the forests
without provision for commercial use, Pinchot favored opening up all the
reserves to scientifically managed economic development. This ideological
divide widened first in 1897 when Pinchot publicly endorsed sheep grazing
in the forest reserves, and became a chasm in 1905 when Pinchot, acting as
newly appointed Forest Service Chief, supported plans to dam Yosemite
National Park's Hetch Hetchy valley to create a reservoir that would quench
the thirst of San Francisco's growing population. While Pinchot argued that
the valley's high elevation in the Sierra mountains made it perfect for supply-
ing both drinking water and electric power, Muir, whose Sierra Club had successfully opposed the Hetch Hetchy dam since San Francisco officials first
proposed it in the 1890s, launched an eight-year nationwide campaign to
protect the valley from development. In one of his most well-known statements regarding the controversy, Muir countered Pinchot's economic argument by emphasizing Hetch Hetchy's noncommercial attributes. "Dam
Hetch Hetchy!", he wrote in 1912, "As well dam for water-tanks the people's
cathedrals and churches, for no holier temple has ever been consecrated by
the heart of man."[40] President Woodrow Wilson's decision in December of

1913 to approve the Hetch Hetchy dam was said to have broken the spirit of John Muir, who died less than a year later. Perhaps more importantly, it splintered what was already a fragile alliance within the conservation movement into conservationist and preservationist camps.[41]

Although Franklin Roosevelt was well aware of the difference between Muir and Pinchot's philosophies, his experiences at Hyde Park and his career as a New York state senator indicated that his thinking corresponded more with that of Gifford Pinchot. Roosevelt's desire to produce timber efficiently both on his Hyde Park estate and in the Adirondacks, as well as his reliance on experts including Nelson Brown of the Syracuse forestry school and Gifford Pinchot of the United States Forest Service, placed him squarely in the conservationist, as opposed to the preservationist, camp. Moreover, Roosevelt's "new theory" of placing community liberty above that of the individual, and his frustration with lumbermen who could not "see more than about six inches in front of their noses," echoed to a great extent Pinchot's philosophy of the "greatest good of the greatest number in the long run." And finally, unlike Muir's desire to preserve natural beauty for spiritual reasons, Roosevelt believed that "aesthetic considerations . . . play a very small—in fact, a negligible part." This is not to say that Roosevelt had no preservationist tendencies. For instance, while scientifically managing his tree plantings at Hyde Park he also ensured that a stand of old-growth forest located along the western edge of his property was left untouched so that it could be "preserved just as nature always has treated it."[42] Although such openness to Muir's philosophy would become increasingly important to the CCC story as the Great Depression wore on, during the Progressive era Roosevelt was clearly a disciple of Pinchot's "gospel of efficiency," instead of Muir's transcendental church.

Franklin Roosevelt's inclination for Pinchot-style conservation was still very much intact when he became governor of New York in 1929, long after progressivism was said to be dead and buried.[43] In his 1931 "Message to the Legislature," for example, Roosevelt noted the alarming rate of farm abandonment throughout the state and proposed a solution that reiterated his utilitarian notion of natural resources. "Every acre of rural land in the state," he explained, "should be used only for that purpose for which it is best fitted and out of which the greatest economic return can be derived."[44] To help formulate this policy, the new governor also looked to experts, much as Progressive era conservationists had. He appointed Henry Morgenthau, editor of the *American Agriculturalist* to the post of New York state conservation commissioner, and relied often on the advice of George Warren and Carl Ladd of Cornell's College of Agriculture, whose land-use studies advocated the removal of marginal and submarginal farmland from production. Yet perhaps most indicative of Roosevelt's persistent belief in Progressive era conservation was his support of the Hewitt Amendment. Introduced by conservative senator Charles Hewitt in 1931, the amendment to New York's constitution authorized the state to purchase abandoned farmland, reforest it, and scientifically manage it as production forests. Roosevelt encouraged Hewitt to introduce the amendment, enlisted his old friend Gifford Pinchot to campaign for its passage, and steadfastly supported it even when Al Smith, who continued to harbor presidential aspirations for 1932, openly criticized the measure as a "tree stealing" program.[45] The amendment's passage in 1931 not only indicated that Progressive era conservation continued to function throughout the 1920s, but it also helped propel Franklin Roosevelt toward the White House. As the *New York Times* reported in a front page article, the Hewitt Amendment represented a "victory for Governor Roosevelt and add[s] to his prestige as the titular leader of his party in the State and, for the moment the leading Democratic aspirant for the presidential nomination."[46]

Thus when Franklin Roosevelt asked Congress to create the CCC in March of 1933, his thinking concerning the New Deal program had been greatly influenced by his involvement in the Progressive era conservation movement. His desire to "create a civilian conservation corps to be used in simple work . . . confining itself to forestry, the prevention of soil erosion, flood control and similar projects" reflected the president's experiences both managing his family's estate in Hyde Park and serving in the New York State Senate, where he maintained close ties to Gifford Pinchot and expressed an affinity for the chief forester's philosophy regarding the efficient production and use of natural resources. Moreover, Roosevelt's actions as governor of the Empire State suggest that his belief in mainstream conservation, as opposed to the more radical preservationist beliefs of John Muir, remained with him even during the late 1920s, when other progressive reform efforts had faded away. Yet the deterioration of the nation's natural resources was not the only crisis worrying Roosevelt when he asked Congress to create the CCC in March of 1933. Unemployment, particularly among young men, also concerned the president. Not surprisingly, Franklin Roosevelt had a long history with youth relief during and after the Progressive era that equally, if not to a greater extent, influenced his thinking concerning the Corps.

Franklin Roosevelt's March 21, 1933 congressional address drew attention to the unemployment problem then facing the nation, and proposed work in the CCC as part of the solution. Roosevelt's language throughout his message, however, suggested that the new president held specific notions concerning both those who were jobless and the type of jobs the Corps would provide. For instance, Roosevelt's address incorrectly suggested that unemployment during the early years of the Great Depression was primarily an urban problem. "The overwhelming majority of unemployed Americans," he explained to Congress in his CCC address, "are now walking the streets."

Two months earlier Roosevelt had made similar remarks, telling a crowd "there are hundreds of thousands of boys who only know the pavements of cities and that means that they can take only those jobs that are directly connected with the pavements of cities."[47] Similarly, immediately after Congress established the CCC, the president invited representatives from 17 of the nation's largest cities to the White House to both explain his plans for employing young men and to ask for their help in recruiting the Corps' first 250,000 enrollees.[48] And whereas the president viewed joblessness as an urban problem, he conversely envisioned its solution as taking place in the countryside, where he planned to locate CCC camps. Through the Corps, he wrote in his congressional message, "we can take a vast army of these unemployed out into healthful surroundings."[49] Thus, while Roosevelt the conservationist saw the countryside's natural resources as sickly and in need of scientific management, Roosevelt the unemployment reformer believed the countryside was also potentially rejuvenative, especially for young urban men.

This belief that the countryside was a curative for urban problems was not new to Roosevelt; it had informed much of his thinking during the early years of the Great Depression. In August of 1931, for instance, Governor Roosevelt gave a speech before the American Country Life Conference in Ithaca, New York, in which he painted a gloomy portrait of urban America. "In times of economic depression we expect to find a concentration of unemployed persons, and as a result a concentration of distress, in the cities," he explained. Rural America, on the other hand, had surpluses of foodstuffs and other benefits as well. "The country has added advantages that the city cannot duplicate in opportunities for healthful and natural living," Roosevelt told his audience in Ithaca. "There is contact with earth and with nature and the restful privilege of getting away from pavements and from noise." As governor, it was Roosevelt's responsibility to correct this demographic imbalance by promoting the migration of unemployed workers from New York's urban areas to its rural regions where the materials for healthful living were cheap and abundant. "The task," he concluded in his speech, "is to determine to what extent and by what means the State and its subdivisions may properly stimulate the movement of city workers to rural homes."[50]

As the Ithaca Country Life Conference indicated, Roosevelt was not alone in juxtaposing rural and urban America. During the early years of the Great Depression, a loose coalition of individuals and groups from across the political spectrum called for the resettlement of unemployed urbanites in the countryside.[51] In the late 1920s Ralph Borsodi, a former advertising executive turned social critic, was the leading advocate of this antiurban, back-to-the-land movement. Borsodi left Manhattan in 1922 and moved to the countryside near Suffern, New York, where he and his family built a stone house on seven acres of land, raised domesticated animals, grew their own food, and made their own clothing and furniture. Borsodi described his experiment in self-sufficiency in two best-selling books, *This Ugly Civilization,* published in 1929, and *Flight from the City,* written four years later.[52] Other liberal writers followed Borsodi's lead, publishing what became a whole genre of do-it-yourself guidebooks devoted to explaining in detail how one could leave urban industrial society and survive on one's own in rural America. Edward Parkinson's "The Retreat from Wall Street" (1931), Louise Owen's "Escape from Babylon" (1932), Katrina Hinck's "A Home for $130" (1933), and Maurice Kains's *Five Acres and Independence* (1935), which advised former clerks and factory workers how to operate a farm, were only a few of the best-known works written during the early years of the Great Depression in response to widespread urban unemployment.[53]

This Depression era back-to-the-land sentiment also gained credibility on the political right from a group of southern intellectuals who became known as the Nashville Agrarians. Centered at Vanderbilt University, members of this informal group included English professors, literary critics, poets, historians, and economists, such as Allen Tate, John Crowe Ransom, Donald Davidson, Frank Owsley, Robert Penn Warren, and Andrew Lytle. The Agrarians' central theme, put forward most succinctly in their 1930 collection of essays titled *I'll Take My Stand,* centered around what they believed to be an unavoidable conflict between industrialism and agrarianism. According to the Vanderbilt intellectuals, industrialism represented everything antitraditional, immoral, and deadening, while agrarianism stood for all that was stable, moral, and spiritually uplifting. Together they viewed urban society as repugnant, and eulogized life in the countryside. "A city of any sort," wrote John Crowe Ransom, "removes men from direct contact with nature, and cannot quite constitute the staple or normal form of life for the citizen."[54] Joining the Agrarians on the right during the early 1930s was the Catholic Rural Life movement. Centered in the Midwest rather than the South and led by Fathers Luigi Ligutti, John Rawe, and W. Howard Bishop, this coalition promoted economic independence and family unity by encouraging widespread ownership of farms. It accomplished this by funneling financial support to rural parishes and by establishing "rural-life bureaus," directed by priests, to facilitate the "colonization" of rural regions. The return of Depression era families to the land, wrote Howard Bishop in *Landward,* the Catholic Rural Life movement's official bulletin, symbolized "the foundation for a real civilization upon which an enduring Christian structure can be found."[55]

The antiurban, pro-rural sentiment expressed by Roosevelt and others during the late 1920s and early 1930s was only the newest incarnation of a much older movement whose ideological roots went back at least to Thomas Jefferson and more recently to Henry David Thoreau. Much of this sentiment crystallized during the Progressive era when concern over the deleterious affects of industrialization, urbanization, and what Alan Trachtenberg has called the "incorporation of America" was at its peak.[56] While some Americans accepted these changes, others resisted by joining unions, embracing populism, and becoming active in a whole host of progressive reform efforts aimed at insulating the working class from urban dangers.

Before the turn of the century these reformers, most of whom hailed from the middle and upper classes, practiced what historian Paul Boyer has termed "coercive moral reform," meaning they attempted to stamp out urban vices such as prostitution, alcohol consumption, and gambling, through moral persuasion and legal repression. During the first decade of the twentieth century, however, a new strategy of social control linked to advances in behavioral psychology emerged to compete with these coercive reform efforts. Called "environmentalism" by its adherents, this novel approach shared the underlying moral assumptions and aims of coercive efforts but differed fundamentally on how to achieve these goals. Instead of overtly repressing urban vices, environmentalist reformers hoped to create an urban setting where objectionable behavior would not be practiced and would thus wither away. Summing up this new philosophy, John Dewey wrote in 1908 that the most effective social control was not based on the legal enforcement of strict behavioral standards, but rather on "the intelligent selection and determination of the environments in which we act."[57] Those involved in environmentalist efforts such as the city beautiful movement, city planning, housing reform, settlement work, and even urban sanitation, thus believed not in good and bad people, but rather in good and bad surroundings.[58] The progressive definition of environmentalism was therefore radically different from the term employed during the post–World War II period to describe the movement that would identify itself with Earth Day.

Progressive era environmentalists believed their reform strategy was particularly effective in influencing young adults, who many viewed as malleable putty ready to be sculpted into model citizens. Landscape designers such as Frederick Law Olmsted, Charles Elliot, and George Kessler, for instance, promoted the city park as a "harmonizing and refining" influence "favorable to courtesy, self-control, and temperance," especially for the city's younger generation.[59] As Kessler put it, the "green turf" and "waving trees" of urban parks produced "innocent, joyous" youngsters instead of "dirty, white-faced, and vicious gamins" prone to "immorality and vice."[60] Another park enthusiast agreed, noting that when city dwellers "have Nature at hand, evil seems weakened . . . the souls of children become freshened with joy."[61] The playground movement, led by Henry Stoddard Curtis of the Playground Association of America, was similarly concerned with youth. In Curtis's 1917 classic, *The Play Movement and Its Significance,* he depicted urban youth as prone to congregate on street corners "where drinking and the sex lure are the main enticements." For Curtis the answer lay not in repressing such activities but in creating "a different environment" as an alternative to the vice-ridden streets. The playground was one such alternative.[62]

Along with creating "different environments" within the metropolis, progressive environmentalists were also interested in the healing potential of the nonurban setting. This belief in the rejuvenative power of the countryside was part of a larger trend that gripped middle- and upper-class Americans during the early decades of the twentieth century. Called "antimodern" by Jackson Lears, "a search for the simple life" by historian David Shi, and a "wilderness cult" by Roderick Nash, this progressive nature craze had three main elements: a country life movement similar to the back-to-the-land sentiment of the early depression years, a wilderness fad that focused on preserving and experiencing life in the wild, and finally an outdoor fresh air movement best exemplified by country vacations and summer camps.[63] The most vociferous promoter of this outdoor rage was Theodore Roosevelt, who during the last decade of the nineteenth century became increasingly convinced that urban American was becoming an "overcivilized man, who has lost the great fighting, masterful virtues." To counter this descent into "flabbiness" and "slothful ease," in 1899 Roosevelt began urging Americans to adopt what he called "the doctrine of the strenuous life," which entailed a "life of toil and effort, of labor and strife."[64] Central to this sort of living was direct contact with nonurban nature.

Progressive environmentalist reformers enthusiastically embraced Theodore Roosevelt's call for a more strenuous life, and went to great lengths to transport young urbanites beyond city limits to more bucolic surroundings. This was especially true during the summer months, when many urban youths were out of school but unable to find employment. Edward Bok, editor of the *Ladies Home Journal,* was a leading proponent of so-called Fresh Air Funds, and helped establish several in eastern cities to subsidize country vacations for out-of-work city-bred adolescents. According to Bok and his associates, the urban environment's filth, its weary working conditions, and its general enervating influence could be overcome by spending a few weeks in the "simplicity and sincerity of nature."[65] Similar beliefs underwrote the youth camping movement, which became a fad in its own right during this period. For those "who cannot afford yachting trips and the like, and whose ideas of summer recreation are not attuned to the string band of a 'summer hotel,' " explained one youth camping advocate, "there is nothing that returns so much for the expenditure of strength and money as plain American camping."[66]

One of the foremost promoters of the youth camping movement during the Progressive era was the Boy Scouts of America. Although Lieutenant

General Sir Robert Baden-Powell officially founded the Boy Scouts in England in 1907, the Boer War hero borrowed heavily from a number of similar organizations already established in the United States. One such group was the Woodcraft Indians, founded by popular nature writer Ernest Thompson Seton. In a series of articles appearing in Edward Bok's *Ladies Home Journal* in 1902, Seton described how his new group organized boys into tribes, taught them games based on Indian legend and ceremony, bestowed awards for good conduct, and most importantly took them out of the city to camp in the countryside. Alongside Seton's Indians, Daniel Beard founded the Sons of Daniel Boone in 1905 as a circulation booster for an outdoor recreation magazine. Yet whereas Seton relied on native American history and symbols to encourage young boys to take to the outdoors, Beard employed the pioneering heritage and focused on teaching outdoor survival techniques such as fire building, map reading, and shelter construction. When a coalition of New York City youth reformers officially established the Boy Scouts of America in 1910, focusing enrollment in the program on adolescents and young men between the ages of nine and twenty, it not only incorporated these organizational forerunners but found jobs for both Seton and Beard, with the former acting as Chief Scout from 1910 to 1915 while Beard served as National Scout Commissioner of the United States from 1910 until his death.[67]

The philosophy of the Boy Scouts of America reflected the beliefs of Progressive era environmentalists, and it is no wonder that scouting was an integral component of their reform efforts.[68] This affinity was perhaps best expressed in the Boy Scouts' first *Handbook,* which Seton wrote in 1910. A century ago, the *Handbook* began, American boys lived close to nature, but since then the nation had experienced "unfortunate change" marked by industrialization and the "growth of immense cities." The result, Seton warned, was "degeneracy" and an urban population that was "strained and broken by the grind of the over-busy world." As with other environmentalist efforts, the Boy Scouts' solution was not to destroy urban dangers but rather to introduce scouts to "outdoor life . . . nearest to the ground" so that these boys could "live the simple life of primitive times." The Scout *Handbook* also provided instruction in woodsmanship and camping in an effort to urge city boys to spend at least one month each year in the countryside away from urban civilization.[69] To foster this, soon after its creation in 1910 the Boy Scouts of America began establishing campgrounds on the outskirts of metropolitan areas throughout the country.

Similar to his cousin Theodore, who became the Boy Scouts' first Chief Scout Citizen in 1912, Franklin Roosevelt was deeply involved in the scouting movement. The younger Roosevelt's experiences with the Boy Scouts began in 1921, when he accepted the chairmanship of the organization's

Greater New York City Council. The following year Roosevelt helped centralize the scouting movement in and around the city by creating the Boy Scout Foundation of Greater New York, which coordinated the work of the five borough councils that had previously maintained independent relationships with the Boy Scouts of America's national leadership. According to Roosevelt, such centralization would put the organization on a more "uniform and practical basis."[70] In 1922 he became president of the Boy Scout Foundation of Greater New York and remained in that position until he resigned in 1937.

Throughout his 16-year involvement with the Boy Scouts, Roosevelt aggressively promoted a progressive environmentalist agenda. This became only too clear at a Boy Scout dinner in March of 1929, during which he emphasized the role of the physical setting in shaping the development of young urbanites. "The records show," Roosevelt told the audience gathered in New York's posh Metropolitan Club, "that the question of environment is important."[71] At a similar Boy Scout event a few years later he expressed the belief, also shared by Progressive era environmentalists, that urban surroundings were particularly dangerous for city youths, especially those from working-class families. For "the city boy living in crowded conditions," Roosevelt explained, "artificial interests have been substituted. Normal, natural growth is threatened."[72] He likewise promoted the other side of the environmentalist coin, namely that the countryside was a potential curative for the problems afflicting urban youth. When transported to the countryside the urban boy "discovers that the woods, the birds, the fields, the streams, the insects speak a language he understands," Roosevelt wrote in a 1928 article on the Boy Scouts for the *New York Times*. "His new environment takes on the aspect of a vast nature-lore museum which beckons to him to enter its great domain of study and to discover for himself."[73]

For the Boy Scouts, and for Franklin Roosevelt in particular, the means of introducing city boys to the benefits of the nonurban environment was through camping in the countryside. When Roosevelt became chairman of the Boy Scout Foundation of Greater New York in 1922, the organization maintained 18 scout campgrounds in the Bear Mountain section of the Palisades Interstate Park, which was located on the western bank of the Hudson River approximately halfway between New York City and Roosevelt's Hyde Park estate. Reputed to be the largest camping facility for scouts in the world, the Bear Mountain camps varied in size and together accommodated, at any given time, approximately 2,200 young men, most of whom spent three weeks in the park. Yet while more than 6,000 boys camped at Bear Mountain during the summer of 1922, two-thirds of New York City's 20,000 scouts were unable to do so because of a lack of camping space.[74] As the Foundation's annual report for 1922 lamented, "thousands of the

New York City Scouts have not had the privilege of attending" the Bear Mountain camps. Franklin Roosevelt was especially concerned, stating years later that "we realized then that our camping facilities were inadequate."[75]

As president of the Boy Scout Foundation of Greater New York, Roosevelt responded to this dearth of camping opportunities for urban unemployed youths by initiating a campaign to increase the number of scout campgrounds throughout the state. He began the effort as early as 1921. "It is probable that next year we will increase our capacity so that we can take care of as many as 3,000 boys daily," wrote Roosevelt to a Boy Scout backer, "and I would not be surprised if within the next three to five years we would have as many as 5,000 to 10,000 boys in camp at one time."[76] During the next several years Roosevelt raised funds in hopes of acquiring land in upstate New York for campground purposes, and in 1929 began developing a 10,600-acre tract located about 100 miles north of New York City in Sullivan County, where he estimated that 100,000 scouts could experience outdoor living each summer.[77] As the Boy Scouts of America's national leadership reported that year, experience in camps such as the one in Sullivan County "includes many outdoor activities which bring Scouts closer to nature."[78]

Just what the Boy Scouts of America meant by "closer to nature" was evident in the daily operation of scout camps nationwide during the early years of the organization's history. With its roots in both Seton's Woodcraft Indians and Beard's Sons of Daniel Boone, the Boy Scouts understandably promoted outdoorsmanship and woodcraft in their campgrounds rather than nature study. Scoutmasters regularly taught the young boys how to pitch tents, blaze trails, and build fires, but rarely how to appreciate or understand the countryside around them. In fact, the Boy Scouts' official camping program included only one test among many that involved the identification of trees and animals.[79] This desire to dominate or subdue nature was perhaps best epitomized by the Boy Scout hatchet, which the organization sold to the boys in such numbers that early on it became a standard addendum to the Scout uniform. The teaching of trailblazing, the use of wood for rustic bridges and fences, and the continual search for firewood in and around Boy Scout campgrounds, all encouraged a slash-and-burn style of outdoor living. Each year, for instance, scouts visiting the Bear Mountain camps stripped bark from birch trees as high as they could reach, often while posing for publicity photographs.[80] Such young men were obviously ignorant of the conservation efforts underway 50 miles upriver on Franklin Roosevelt's Hyde Park estate, or 100 miles north in the Adirondack Forest Preserve.

The Boy Scouts of America thus had much in common with Franklin Roosevelt's CCC. Both were concerned with unemployed male youths and associated this problem with the urban setting. Each also promoted the relocation of young city dwellers to camps in the countryside as a curative.

In doing so, the Boy Scouts and the Corps were only the most recent in a long line of environmentalist efforts begun by Frederick Law Olmsted, Theodore Roosevelt, and Edward Bok, which continued during the early years of the Great Depression by such back-to-the-landers as Ralph Borsodi, the Nashville Agrarians, and Father Howard Bishop. Yet as the Boy Scouts' camping program indicates, the conservation of natural resources was not an integral component of the organization's philosophy. Rather than teaching their charges how to efficiently use timber, soil, and water, scoutmasters allowed, and often promoted, the waste of such resources. How, then, did the idea to conserve natural resources become linked, in Franklin Roosevelt's mind to the notion of rejuvenating urban male youths?

Franklin Roosevelt first suggested the idea of introducing the ideology and practice of conservation to scouting soon after he became president of the Boy Scout Foundation of Greater New York in 1922. In a letter that year to Foundation member George Pratt of Brooklyn, Roosevelt expressed his desire to correct the Boy Scout camping program's wasteful natural resource practices. "I shall do everything possible," explained Roosevelt of his new duties as Foundation president, "to expand what might be called the better understanding of nature by these city-bred boys."[81] That same year Roosevelt asked members of the Boy Scout camping committee to investigate the possibility of enlarging the study of forestry at the scout camps already in existence. He also asked the Palisades Interstate Park Commission if it would be possible to secure a small tract of land near the Boy Scout camps in the Bear Mountain section of the park "to be scientifically forested by the boys this summer."[82] The response of the Palisades Park Commissioners fit well with Roosevelt's conservationist ideology, which as noted above was centered around the desire to efficiently use and develop natural resources. Such a reforested tract, the Commission replied, would "imbue the coming generations with a knowledge of forest conservation" and "[teach] business, not sentimental theory."[83] It would seem that as soon as Franklin Roosevelt became involved with the Scouts, the days of the boys' overactive hatchet were numbered.

Roosevelt expanded his effort to promote conservation within the Boy Scouts in 1923. Rather than merely adding forestry study to the camping program already in existence, Roosevelt helped establish new scout camps, also in the Palisades Interstate Park, dedicated specifically to the teaching and practice of natural resource conservation. The Boy Scout Foundation of Greater New York established the first of such camps in 1923, and another the following year. Known collectively as the "Franklin D. Roosevelt Conservation Camps," each could accommodate approximately 60 campers and accepted only older scouts no less than 15 years of age, many of whom had trouble finding employment during the summer months.[84] Instead of pitching tents, building fires, and stripping birch bark, as was taught at the regular scout campgrounds, boys attending these special conservation camps would perform work such as cutting fire breaks, fighting forest fires, and planting trees. As one forester involved in the new program put it, the conservation camps would "appeal to the older scout," "give him expert training . . . and familiarize him with one of the biggest economic problems of the day—Forestry."[85] In 1929 Roosevelt decided to expand the scout conservation program yet again by including forestry work in the development plans for the Boy Scout Foundation's new 10,600-acre campground in Sullivan County, New York.

During the early years of the Great Depression, there were signs that the conservation program that Roosevelt initiated in the Palisades camps was influencing Boy Scouts throughout the nation. In 1930, for instance, the Boy Scouts of America launched a five-year reforestation program known as the "Nut Seed and Tree Planting Project." After gathering nuts from trees in Mount Vernon's Arlington National Cemetery and at Theodore Roosevelt's grave in Oyster Bay, Long Island, the organization dispersed them to scouts throughout the country for planting as memorial trees. Boy Scouts in Emmet County, Iowa conducted the project's first tree-planting demonstration, and other troops followed, with scouts planting nearly 50,000 seedlings in Oshkosh, Wisconsin, 20,000 in Leominster, Massachusetts, and 4,700 in Massillon, Ohio. Troops from Roosevelt's home county of Dutchess, New York, planted 4,500 seedlings and received in return "that satisfaction that comes from planting trees."[86] All told, the Boy Scouts of America hoped to plant more than five million trees during the five-year project. More importantly, the conservation of natural resources had finally been wedded to environmentalist youth reform in the mind of Franklin Roosevelt.

When the stock market crashed in October of 1929, Governor Roosevelt applied what he had accomplished with the Boy Scouts to the economic and environmental problems of the Empire State. Just a few months after black Tuesday, Roosevelt asked the state legislature for an appropriation to fund a tree-planting program, similar to that initiated by the Boy Scouts, to provide jobs for New York's growing unemployed population.[87] In August of 1931 the governor greatly expanded this sort of relief work when he established the Temporary Emergency Relief Administration (TERA), which along with providing food, clothing, and shelter to those in need, also created jobs, many in the field of forestry. As already noted, earlier that year New Yorkers had voted in favor of the Hewitt Amendment, which authorized the state to purchase abandoned farmland for reforestation purposes. The forestry jobs created by TERA provided the labor necessary to physically convert these farmlands into forests. Roosevelt named Harry Hopkins, former head of New York's Tuberculosis and Health Association, executive director of TERA and directed him to coordinate the program's forestry work with Conservation Department commissioner Henry Morgenthau, Jr.[88]

Under Roosevelt's close supervision, TERA began its forestry relief work early in 1932. "We have lately undertaken a new project using our State relief funds in forestry work," explained the New York governor to Ovid Butler, executive secretary of the American Forestry Association. "On this project we are now employing 100 men in Central New York on a somewhat experimental basis to find out to what extent we can profitably use men from the lists of the unemployed to improve our existing reforestation areas."[89] The "experiment," as Roosevelt called it, proved so successful that TERA immediately created more than 10,000 conservation-related jobs for out-of-work New Yorkers.[90] Not surprising given Roosevelt's contemporaneous involvement with scouting, TERA's forestry work initiative shared numerous traits with the Boy Scout Foundation of Greater New York's conservation program. For example, TERA laborers performed work such as the cutting of fire lanes, the clearing of dead wood, and the planting of seedlings, that was nearly identical to that done by Boy Scouts in the Franklin D. Roosevelt Conservation Camps. Moreover, TERA conducted much of this work in the Palisades Interstate Park, literally a stone's throw from the Boy Scouts' conservation campgrounds. All told, by the end of 1932 Roosevelt's forestry work-relief program was aiding more than 25,000 unemployed New Yorkers.[91]

The forestry relief work undertaken by TERA not only reflected the practices of Franklin Roosevelt's Boy Scouts of Greater New York, but also foreshadowed the birth of the CCC. For instance, both TERA and the CCC housed their workers in camps located on public lands, and each program provided its participants with food, shelter, and an allowance in exchange for labor. In TERA's case this pay came to 12 dollars a month, about half what the Corps provided. TERA also acted as selection agent for CCC enrollees in New York state from 1933 until 1937.[92] Most importantly, however, the TERA conservation program soon went national, much as the CCC would in March of 1933. Although President Herbert Hoover showed his affinity for Governor Roosevelt's forestry relief initiative by allocating funds for similar work in National Forests in January of 1931, it was in November of that year that state officials from across the country began praising New York's TERA.[93] "I saw in the paper your plan . . . in regard to relieving unemployment and perpetuating forest growth," wrote an admirer from Mississippi to Roosevelt, "and I wish to take this opportunity to congratulate you on this wonderful forward advancement and I am hoping that our state and a great many other states will follow in your footsteps."[94] Numerous states did. The Great Lakes states of Michigan and Indiana, others in the south such as Mississippi and Virginia, and Oregon and Washington state in the pacific northwest, among others, all established forestry work-relief programs based largely on TERA.[95] Even California's forestry relief efforts, often cited as the most advanced program in the country, began six months after Roosevelt asked the New York state legislature to establish TERA. In addition, when the Golden State's forestry relief camps began closing in May of 1933, it was the CCC camps that replaced them.[96] Thus rather than being the ideological source of the Corps, as many have posited, TERA was more an intermediary step on the state level between Roosevelt's municipal Boy Scouts conservation initiative and his federal CCC program.

As Franklin Roosevelt's pre-presidential experiences indicate, the Corps idea honored by the nation's first CCC enrollees, who named their camp near Luray, Virginia "Camp Roosevelt," was indeed a conflux of desire and need. Roosevelt's desire to conserve the nation's natural resources in light of the disasters occurring along the Ohio and other rivers was the first tributary in the ideological stream that ultimately led to the creation of the CCC. This desire was influenced by the future president's involvement in the Progressive era conservation movement both while managing his family estate in Hyde Park, and while acting as New York state senator and governor in Albany. Roosevelt's conservationist ideology was likewise shaped by Gifford Pinchot, chief forester of the U.S. Forest Service. Unlike John Muir, who believed in preserving nature for its aesthetic and spiritual properties, Roosevelt and Pinchot promoted the rational development and use of natural resources through scientific management. The work performed by the CCC, which Roosevelt defined narrowly in his congressional address as "forestry, the prevention of soil erosion, flood control and similar projects," was thus partly the offspring of the conservation movement's preoccupation with rational production.[97]

This conservationist stream, however, accounts for only one tributary flowing into the CCC's ideological origin. Franklin Roosevelt's experiences with youth relief also shaped the Corps's genesis. Through his involvement with the Boy Scout Foundation of Greater New York, Roosevelt came into contact with reformers who like himself viewed the urban setting as threatening and the countryside as potentially rejuvenative. Similar to Frederick Law Olmsted before them, progressives including Ernest Thompson Seton, Edward Bok, and Theodore Roosevelt, all proposed strenuous stints in the great outdoors as a curative for city ills. Franklin Roosevelt made such experiences a reality for tens of thousands of New York City adolescents by increasing the number of Boy Scout campgrounds in the metropolitan region. He did the same when asking Congress in March of 1933 to create a conservation corps that would "take a vast army of these unemployed out into healthful surroundings." Along with the Progressive era conservation movement, then, President Roosevelt's thinking concerning the CCC was also shaped by progressive environmentalist reform.

In relying on his past experiences with both conservation and progressive environmentalism, Franklin Roosevelt was more of a synthesizer of existing ideologies than a creator of new ones. The president was original, however, in his desire to combine the conservation of natural resources with the conservation of young men. He initiated this process by establishing the Franklin D. Roosevelt Conservation Camps in New York's Palisades Interstate Park, and furthered it when as governor of New York he created forestry relief work under TERA. Because of such actions, in August of 1933 the newly elected president could confidently announce to thousands of cheering scouts that "this Spring, because of my scout training, I took a leaf out of the notebook of scouting" and "started the CCC in this country, modeling it to a large extent after scouting."[98]

The conflux of desire and need resulting in the Corps had implication for the conservation movement during and after the Great Depression. As stated above, during the first two decades of the twentieth century the conservation movement experienced internal divisions that pitted amateur preservationists such as John Muir against professional conservationists led by Gifford Pinchot. Muir's spiritual wilderness sat in direct opposition to Pinchot's utilitarian timber. Yet in creating the Corps, Franklin Roosevelt altered this equation by adding environmentalist thinking to this conservationist–preservationist paradigm. As a result no longer would Gifford Pinchot and the ghost of John Muir battle alone over conservation controversies like Hetch Hetchy. Instead during the New Deal era, conservation and preservation would be joined by a third ideology, promoted by the likes of Frederick Law Olmsted and the Boy Scouts, which viewed Nature as a healthful rejuvenator of urban youths. This intellectual addition would help transform the conservation movement after World War II in ways that fostered environmentalism.

NOTES

1. The naming of "Camp Roosevelt" and the construction of the "Camp Roosevelt" sign is mentioned in William Train, Jr., "Building Camp Roosevelt . . . The First Civilian Conservation Corps Camp in the U.S.," *National Association of CCC Alumni Journal* (hereafter NACCCA) (November 1992): 6–7. For a general description of the establishment of the first CCC camp see, Erle Kauffman, "Roosevelt—Forest Camp No. 1," *American Forests* (June 1933): 251; Captain Leo Donovan, "The Establishment of the First Civilian Conservation Corps Camp," *Infantry Journal* (July–August 1933): 245; Emergency Conservation Work, Office of the Director, Washington, DC, "Memorandum for the Press, Release to Morning Papers, Friday, April 17, 1936," Printed Materials Collection, Civilian Conservation Corps, Press Releases, Franklin Delano Roosevelt Library (hereafter FDRL), 1; and Alison Otis, *The Forest Service and the Civilian Conservation Corps: 1933–1942*

(Washington, DC: United States Department of Agriculture, U.S. Government Printing Office, 1986), 84.

2. Baker makes his claim in his book *Green Glory* (New York: A.A. Wyn, 1949), 66–68. Louis M. Howe to Isabel Erlich, October 4, 1933, Official File 268, Folder "Miscellaneous Sept–Oct 1933," FDRL; Robert Fechner to James Farley, January 19, 1937, Official File 268, Folder "Jan–Feb 1937," FDRL.

3. Depression era accounts that link James's essay to the origin of the CCC include: Captain X, "A Civilian Army in the Woods," *Harpers* (March 1934), 487; F.A. Silcox, "Our Adventures in Conservation," *Atlantic Monthly* (November 1937), 714; "Conservation: Poor Young Men," *Time,* February 6, 1939, 10; and Kenneth Holland and Frank Hill, *Youth in the CCC* (Washington, DC: American Council on Education, 1942), 16; William James, "The Moral Equivalent of War," *McClures* 35 (1910): 463–68.

4. For a good account of European youth work programs prior to the creation of the CCC see, Holland and Hill, *Youth in the CCC,* 19. On Bulgarian and Swiss youth work programs also see, Silcox, "Adventures in Conservation," 714; For programs in Holland see, Arthur Ringland, "Land Utilization and the Unemployed in Holland," *American Forests* (August 1932): 448. And for Germany see, Arthur Ringland, "The CCC in Germany," *Journal of Forestry* 34 (1936): 554–61. See also, "CCC," *Life* (June 1938), 58.

5. Jonathan Mitchell, "Roosevelt's Tree Army: I," *New Republic,* May 29, 1935, 64. Also see, Silcox, "Our Adventures in Conservation," 717; and "200,000 Wandering Boys," *Fortune,* February 1933, 47. Of the two million Americans drifting from town to town on freight cars during the early years of the Great Depression, approximately 250,000 were young people. John Salmond, *The Civilian Conservation Corps, 1933–1942: A New Deal Case Study* (Durham, NC: Duke University Press, 1967), 3.

6. For an account of Roosevelt's experiences on his Hyde Park estate see, Nelson Brown, "President has Long Practiced Forestry," *New York Times,* April 30, 1933, sec viii, p. 1. On his experiences as New York state senator and governor see "Conservation," *Time,* February 6, 1939, 10.

7. Franklin Roosevelt to Louis Howe, October 10, 1933, as reprinted in Edgar Nixon, comp. and ed., *Franklin D. Roosevelt and Conservation, 1911–1945* (Hyde Park, New York: FDRL, 1957), 1:143. Nixon has collected many of Roosevelt's correspondences and speeches regarding the conservation of natural resources in this two-volume set.

8. Marguerite LeHand to I. Van Meter, July 15, 1939, as reprinted in Nixon, *FDR and Conservation,* 2: 354.

9. The historiography on the CCC is scant. Scholars who have refrained from examining the ideological origins of the CCC include: John Paige, *The Civilian Conservation Corps and the National Park Service: An Administrative History* (Washington, DC: National Park Service, United States Department of the Interior, 1985), 1; Thomas Cox, *This Well-Wooded Land: Americans and Their Forests From Colonial Times to the Present* (Lincoln: University of Nebraska Press, 1985), 218; Stephen Fox, *The American Conservation Movement: John Muir and His Legacy* (Madison: University of Wisconsin Press, 1981), 189;

and John Mitchell, "FDR's Tree Army," *Audubon* (November 1983), 84. The few historians who do delve more deeply into the intellectual roots of the program tend to overemphasize Roosevelt's lifelong involvement in the conservation movement. See especially, Salmond, *The CCC,* 6–7; Barrett Potter, "The Civilian Conservation Corps in New York State: Its Social and Political Impact" (Ph.D. diss., State University of New York, Buffalo, 1973), 7; and Henry Clepper, "The Birth of the CCC," *American Forests* (March 1973): 9.

10. A number of environmental historians have called for more scholarship that illustrates the interrelation of environmental, intellectual, and socioeconomic change. See especially, Arthur McEvoy, "Towards an Interactive Theory of Nature and Culture: Ecology, Production, and Cognition in the California Fishing Industry," *Environmental Review* 11, no. 4 (Winter 1987): 289–305; Donald Worster, "Transformations of the Earth: Toward an Agroecological Perspective in History," *Journal of American History* 76 (1990): 1087–1105. According to environmental historian William Cronon, environmental historians have generally written studies that either link environmental and economic change, or studies that connect intellectual and environmental changes. "Too rarely," he writes "have we had the three together." This dissertation strives to partially fill this historiographical gap. See, William Cronon, "Modes of Prophecy and Production: Placing Nature in History," *Journal of American History* 76 (1990): 1123. On the rise of the modern welfare state see, Meg Jacobs, "How About Some Meat?: The Office of Price Administration, Consumption Politics, and State Building from the Bottom Up, 1941–1946," *Journal of American History,* 84, no. 3 (December 1997): 912. For a social scientistic view of state development, see Margaret Weir, Ann Shola Orloff, and Theda Skocpol, eds., *The Politics of Social Policy in the United States* (Princeton: Princeton University Press, 1988); and Stephen Skowronek, *Building a New American State: The Expansion of National Administrative Capacities, 1877–1920* (New York: Cambridge University Press, 1982). Historians who have recently taken up this issue include, Liz Cohen, *Making a New Deal: Industrial Workers in Chicago, 1919–1939* (New York: Cambridge University Press, 1990); Roy Rosenzweig, *Eight Hours for What We Will: Workers and Leisure in an Industrial City, 1870–1920* (New York: Cambridge University Press, 1983); and Dana Frank, *Purchasing Power: Consumer Organizing, Gender, and the Seattle Labor Movement, 1919–1929* (New York: Cambridge University Press, 1994).

11. For a more detailed account of the events of these weeks, see Salmond, *The CCC,* 10–12.

12. The other two legislative initiatives proposed in Roosevelt's March 21, 1933 congressional address would grant aid to states for relief work and establish a broad public works labor-creating program similar to the Works Progress Administration. The address is reprinted in its entirety in Nixon, *FDR and Conservation,* 1: 143.

13. As reprinted in Nixon, *FDR and Conservation,* 1: 143–44.

14. See Samuel Hays, *Conservation and the Gospel of Efficiency: The Progressive Conservation Movement, 1890–1920* (Cambridge: Harvard University Press, 1959; repr., New York: Atheneum, 1974), 1–4.

15. Because scientific professionals were central to the Progressive era conservation movement, reformers rarely if ever recommended conservation work as a curative for the unemployment problems of untrained youths. On the elite character of the progressive conservation, see Hays, *Conservation and the Gospel of Efficiency,* 2–3; Fox, *The American Conservation Movement,* 110; and Robert Gottlieb, *Forcing the Spring: The Transformation of the American Environmental Movement* (Washington, DC: Island Press, 1993), 56.

16. Here and throughout this chapter I am defining the term "ideology" broadly as "the body of ideas reflecting the social needs of an individual," in this case Franklin Roosevelt. See, *The American Heritage Dictionary,* William Morris, ed. (Boston: Houghton Mifflin Company, 1982), 655.

17. Cox, *Well-Wooded Land,* 215.

18. Franklin Roosevelt to Hendrik William Van Loon, February 2, 1937, President's Personal File, FDRL.

19. Franklin Roosevelt, Speech given at Clarksburg, West Virginia, October 29, 1944, as reprinted in Nixon, *FDR and Conservation,* 2: 603. See also, Fox, *The American Conservation Movement,* 185.

20. As quoted in Thomas Patton, "Franklin Roosevelt and ESF—Training a Forester–President," *Alumni Newsletter, SUNY College of Environmental Science and Forestry,* vol. 83 (Winter 1984), 6. Also quoted in Cox, *This Well-Wooded Land,* 215.

21. Roosevelt planted trees at Hyde Park every year except five, 1919–1923, during which he was serving as secretary of the Navy in Washington or recovering from polio. For good accounts of FDR's tree-planting efforts at Hyde Park, see Nelson Brown, "The President Practices Forestry," *Journal of Forestry* 41, no. 2 (February 1943); Nelson Brown, "President Has Long Practiced Forestry," *New York Times,* April 30, 1933, sec. vii, 1; and Patton, "Franklin Roosevelt and ESF," 397–98.

22. Roosevelt mentions the practice of listing his occupation as a "tree grower" when voting in the (Poughkeepsie, N.Y.) Eagle News, November 7, 1944, 4. Also see, Fox, *The American Conservation Movement,* 185; and F. Kennon Moody, "FDR and His Neighbors: A Study of the Relationship Between Franklin D. Roosevelt and Residents of Dutchess County" (Ph.D. diss., State University of New York, Albany, 1981), 99–100.

23. For a description of FDR's Christmas tree-planting efforts, see Nelson Brown, "The President's Christmas Trees," *American Forests* (December 1941): 552; and Anna Riesch-Owen, *Conservation Under Franklin D. Roosevelt* (New York: Praeger Publishing, 1983), 6.

24. Nelson Brown, "The Presidency Practices Forestry," 93; and Patton, "Forestry and Politics," 398.

25. Brown, "The Presidency Practices Forestry," 93.

26. "Resolutions Proposed by the Hon. Franklin D. Roosevelt in the State Senate of New York During the Legislative Session of 1912," FDR: Papers as New York Senator, "Pamphlet File," FDRL. See also, Fox, *John Muir and His Legacy,* 186.

27. For a good history of the Adirondack Forest Preserve, see Philip Terrie, *Forever Wild: A Cultural History of Wilderness in the Adirondacks* (Philadelphia: Temple

University Press, 1985; repr., Syracuse: Syracuse University Press, 1994), chpt. V. For an account of the Roosevelt–Jones Bill, see Nixon, *FDR and Conservation,* vol. 1, 11–20.

28. Franklin Roosevelt to Dexter Blagden, February 21, 1912, Papers as State Senator, Legislative Files, File 26: Conservation Bills, Jan-Feb 1912, FDRL.

29. According to historian Stephen Fox, Roosevelt referred to Pinchot's slides 30 years after this event. Fox, *John Muir and His Legacy,* 185.

30. The text of the address located at the FDRL is somewhat different than that reported in the Troy Record. See Franklin Roosevelt, "Address before the People's Forum, Troy, New York," March 3, 1912, FDR: Papers as New York Senator, General Subject File, Folder: Speeches by FDR, Oct 1910–May 1912, FDRL; and *The Troy Record,* March 4, 1912.

31. Hays, *Conservation and the Gospel of Efficiency,* 3. Other works on the progressive conservation movement include: Fox, *The American Conservation Movement,* 107–08; James Penick, Jr., "The Progressives and the Environment: Three Themes from the First Conservation Movement," in *The Progressive Era,* Lewis Gould ed. (Syracuse: Syracuse University Press, 1974); Donald Worster, *Nature's Economy: A History of Ecological Ideas* (New York: Cambridge University Press, 1977); Roderick Nash, *Wilderness and the American Mind* (New Haven: Yale University Press, 1967); Gottlieb, *Forcing the Spring;* Michael Williams, *Americans and Their Forests: A Historical Geography* (New York: Cambridge University Press, 1989); and Samuel Hays, *Beauty, Health, and Permanence: Environmental Politics in the United States, 1955–1985* (New York: Cambridge University Press, 1987), 13.

32. Richard White, "American Environmental History: The Development of a New Historical Field," *Pacific Historical Review* 54, no. 3 (August 1985): 299.

33. Because of its elite character, the conservation movement reflects more the model of the Progressive era espoused by historians such as James Weinstein and Gabriel Kolko rather than that of scholars including Robert Wiebe and Richard Hofstadter. See, James Weinstein, *The Corporate Ideal in the Liberal State, 1900–1918* (Boston: Beacon Press, 1968); Gabriel Kolko, *The Triumph of Conservatism: A Reinterpretation of American History, 1900–1916* (New York: Free Press, 1963); Robert Wiebe, *The Search for Order, 1877–1920* (New York: Hill and Wang, 1967); and Richard Hofstadter, *The Age of Reform* (New York: Knopf, 1955).

34. As quoted in Worster, *Nature's Economy,* 266.

35. Gifford Pinchot, *Breaking New Ground* (1974, repr.; Seattle: University of Washington Press, 1972), 359, as quoted in Harold Steen, *The U.S. Forest Service: A History* (Seattle: University of Washington Press, 1976), 75.

36. Historian Donald Worster has called Pinchot "the major architect of the Progressive conservation ideology." Donald Worster, *Nature's Economy,* 266. For more on Pinchot's conservationist thinking see, Paul Hirt, *A Conspiracy of Optimism: Management of the National Forests Since World War II* (Lincoln: University of Nebraska Press, 1999); Michael Williams, *Americans and Their Forests,* 416–22; Hays, *Gospel of Efficiency,* 28–30; Fox, *John Muir and His Legacy,* 111; and Nash, *Wilderness and the American Mind,* 134–38.

37. John Muir, *Our National Parks* (Boston, 1901), 74; John Muir, *My First Summer in the Sierra* (Boston, 1911), 211; John Muir, *John of the Mountains: The Unpublished Journals of John Muir,* Linnie Marsh Wolfe, ed. (Boston, 1938), 138; as quoted in Nash, *Wilderness and the American Mind,* 125.

38. Muir biographer Stephen Fox traces this "amateur" tradition throughout the history of the American Conservation Movement. See especially Fox, *The American Conservation Movement,* chpt. 10. Environmental historians disagree over Muir's relationship to the conservation movement of Gifford Pinchot. Some scholars, such as Samuel Hays, do not include Muir in their examination of Progressive era conservation while others, including Stephen Fox, portray Muir as the driving force of the movement. This dissertation portrays Muir's beliefs as a minority voice, albeit one growing in strength during the New Deal era, within the conservation movement.

39. Pinchot, *Breaking New Ground,* 103, as quoted in Fox, *The American Conservation Movement,* 112. As late as 1901 Muir was writing that "state woodlands [should] not be allowed to lie idle," but are made to "produce as much timber as is possible without spoiling them." John Muir, *Our National Parks,* 363.

40. John Muir, *The Yosemite* (New York, 1912), 261–62, as quoted in Nash, *Wilderness and the American Mind,* 168.

41. Numerous environmental historians have identified the Hetch Hetchy controversy as a critical event in the splintering of the conservation movement into conservationist and preservationist camps. See especially, Fox, *The American Conservation Movement,* 111–13; Nash, *Wilderness and the American Mind,* 135–38; Penick, "The Progressives and the Environment," 125–26; Williams, *Americans and Their Forests,* 413–14 and 456; and Gottlieb, *Forcing the Spring,* 24–28.

42. As quoted in Patton, *Forestry and Politics,* 398.

43. The great majority of scholarship on progressivism defines the Progressive era as drawing to a close sometime around 1920. For a discussion of this periodization, see Daniel Rodgers, "In Search of Progressivism," *Reviews in American History* (December 1982): 113.

44. Franklin Roosevelt, "Message to the Legislature," January 26, 1931, as reprinted in Nixon, *FDR and Conservation,* 1: 79.

45. For good accounts of the political battle over the Hewitt Amendment see, Patton, "Forestry and Politics," 407–15; Bernard Bellush, *Franklin D. Roosevelt as Governor of New York* (New York: Columbia University Press, 1955) 95–98; Cox, *This Well-Wooded Land,* 217; "Roosevelt Pleads For Reforestation," *New York Times,* February 14, 1931, 4; "Smith Pushes Fight on Forest Measure," *New York Times,* October 31, 1931, 3.

46. "Forest Measure Approved: Gov. Roosevelt Adds to His Prestige in Clash with Smith," *New York Times,* November 4, 1931, 1. Also see, "Roosevelt Hailed for Polls Victory: Friends All Over Country Congratulate Him on Vote for Forest Amendment," *New York Times,* November 5, 1931, 2.

47. Franklin Roosevelt Speech File, #600, FDRL; also printed as "Return of Jobless From City to Farm is Roosevelt's Aim," *New York Times,* January 17, 1933.

48. The meeting was held on April 5, 1933. Perry Merrill, *Roosevelt's Forest Army: A History of the Civilian Conservation Corps, 1933–1942* (Montpelier, Vermont, 1981), 11.

49. As reprinted in Nixon, *FDR and Conservation*, 1: 143. Roosevelt's incorrect assessment of the unemployment situation is also discussed in Frank Freidel, *Franklin D. Roosevelt: Launching the New Deal* (Boston: Little Brown and Company, 1973), 80.

50. Franklin Roosevelt, "Speech Before the American Country Life Conference, Ithaca, August 19, 1931," Speech File, #437, FDRL. Also see, Franklin Roosevelt, "Back to the Land," *Review of Reviews* 84 (October 1931): 63–64.

51. There is quite an extensive literature on back-to-the-land movements during the first half of the twentieth century. See especially, David Shi, *The Simple Life: Plain Living and High Thinking in American Culture* (New York: Oxford University Press, 1985). See also, Paul Conkin, *Tomorrow A New World: The New Deal Community Program* (Ithaca, New York: American Historical Association, 1959); David Danbom, "Romantic Agrarianism in Twentieth-Century America," *Agricultural History* 65, no. 4 (Fall 1991); Richard White, "Poor Men on Poor Lands: The Back-to-the-Land Movement of the Early Twentieth Century—A Case Study," *Pacific Historical Review;* and Blaine Brownell, "The Agrarian and Urban Ideals: Environmental Images in Modern America," *Journal of Popular Culture* 5 (Winter 1971).

52. Shi, *The Simple Life,* 227.

53. Danbom, "Romantic Agrarianism," 6.

54. John Crowe Ransom, "The Aesthetic of Regionalism," *American Review* II (January 1934), 306, as quoted in Idus Newby, "The Southern Agrarians: A View After Thirty Years," *Agricultural History* 148. Other sources on the Nashville Agrarians include, Paul Conkin, *The Southern Agrarians* (Knoxville: University of Tennessee Press, 1988); and William Havard and Walter Sullivan, eds., *A Gang of Prophets: The Vanderbilt Agrarians After Fifty Years* (Baton Rouge: Louisiana State University, 1982).

55. Howard Bishop, *Landward* I (Spring 1933), 3, as quoted in Christopher Kauffman, "W. Howard Bishop, President of the National Catholic Rural Life Conference, 1928–1934," *U.S. Catholic Historian* 8, no. 3 (1989): 138. Along with Kauffman, other sources on the Catholic Rural Life movement include, David O'Brien, *American Catholics and Social Reform: The New Deal Years* (New York: 1968); Edward Shapiro, "Catholic Agrarian Thought and the New Deal," *The Catholic Historical Review* 65 (October 1979); and Edward Shapiro, "Catholic Rural Life and the New Deal Farm Program," *American Benedictine Review* 28 (September 1977).

56. Alan Trachtenberg, *The Incorporation of America: Culture and Society in the Gilded Age* (New York: Hill and Wang, 1982).

57. John Dewey, "Intelligence and Morals" (1908), in John Dewey, *The Influence of Darwin on Philosophy* (New York: Henry Holt, 1910), 74, as quoted in Boyer, *Urban Masses,* 225.

58. On the shift from "coercive moral reform" to "environmentalist reform," see Paul Boyer, *Urban Masses and Moral Order in America, 1820–1920*

(Cambridge: Harvard University Press, 1978), chpts. 15–16, especially pp. 220–21. For more detailed accounts of various environmentalist reform efforts during the Progressive era see the following: For advances in behavioral psychology see John Burnham, "Psychiatry, Psychology, and the Progressive Movement," *American Quarterly* 12, no. 4 (1960); On the city beautiful movement, see William Wilson, *The City Beautiful Movement* (Baltimore: Johns Hopkins University Press, 1989); On settlement house reform see Gottlieb, *Forcing the Spring;* On the urban sanitation movement see Martin Melosi, " 'Out of Site, Out of Mind': The Environment and Disposal of Municipal Refuse, 1860–1920," *Historian* 35, no. 4 (August 1973); and on women's environmentalist efforts see Maureen Flanagan, "The City Profitable, The City Livable: Environmental Policy, Gender, and Power in Chicago in the 1910s," *Journal of Urban History* 22, no. 2 (January 1996).

59. Frederick Law Olmsted, "Public Parks and the Enlargement of Towns," 76; as quoted in Boyer, *Urban Masses,* 238. For a good analysis of Olmsted's environmentalist philosophy and his influence on the city beautiful movement see Wilson, *The City Beautiful Movement,* chpt. 1.

60. George Kessler, *Report of the Board of Park and Boulevard Commissioners of Kansas City, Missouri* (Kansas City: Hudson Kimberely, 1893), excerpted in Charles Glabb, ed., *The American City: A Documentary History* (Homewood, Illinois: Dorsey Press, 1963), as quoted in Boyer, *Urban Masses,* 239.

61. G. Washington Eggleston, "A Plea for More Parks, and the Preservation of the Sublimities of Nature," as quoted in Boyer, *Urban Masses,* 239.

62. Henry Curtis, *Play Movement and Its Significance* (New York: Macmillan, 1917), 6, 8, 36, 119; as quoted in Boyer, *Urban Masses,* 244. For a good overview of the progressive playground movement, see Dominick Cavallo, *Muscles and Morals: Organized Playgrounds and Urban Reform, 1880–1920* (Philadelphia: University of Pennsylvania Press, 1981).

63. For a general overview of this nature craze and its three main elements, see Shi, *The Simple Life,* 194. On the Country Life Movement, also see William Bowers, *The Country Life Movement in America, 1900–1920* (Port Washington, New York: Kennikat Press, 1974), 45. For an examination of the wilderness cult during the early years of the twentieth century, see Nash, *Wilderness and the American Mind,* chpt. 9. And for analysis of the outdoor fresh air movement, see David MacLeod, *Building Character in the American Boy: The Boy Scouts, YMCA, and Their Forerunners, 1870–1920* (Madison: University of Wisconsin Press, 1983). T.J. Jackson Lears calls this movement "antimodern" in, T.J. Jackson Lears, *No Place of Grace: Antimodernism and the Transformation of American Culture, 1880–1920* (New York: Panthenon, 1981).

64. Theodore Roosevelt, *The Strenuous Life: Essays and Addresses* (New York: 1905), 7–8, as quoted in Shi, *The Simple Life,* 201.

65. "Camping in the Woodland," *New England Magazine* 18 (1898), as quoted in Shi, *The Simple Life,* 205.

66. Ernest Ingersoll, "Practical Camping," *Outlook* 56 (1897), 324, as quoted in Shi, *The Simple Life,* 206.

67. On the founding of the Boy Scouts, see Macleod, *Building Character,*
    131–32; Shi, *The Simple Life,* 208; and John Dean, "Scouting in America:
    1910–1990," (Ph.D. diss., University of South Carolina, 1992). For addi-
    tional information on the Boy Scouts of America, see David Macleod, "Act
    Your Age: Boyhood, Adolescence, and the Rise of the Boy Scouts of America,"
    Journal of Social History 16, no. 2 (1983); David Shi, "Ernest Thompson
    Seton and the Boy Scouts: A Moral Equivalent of War?," *South Atlantic
    Quarterly* 84, no. 4 (1985); Carolyn Wagner, "The Boy Scouts of America: A
    Model and Mirror of American Society" (Ph.D. diss., Johns Hopkins
    University, 1979); Jeffrey Hanover, "The Boy Scouts and the Validation of
    Masculinity", *Journal of Social Issues* 34, no. 1 (1978); Nash, *Wilderness and
    the American Mind,* 147–49; and on the Boy Scouts in Great Britain, see
    Michael Rosenthal, *The Character Factory: Baden-Powell and the Origins of the
    Boy Scout Movement* (New York: Pantheon Books, 1984).

68. According to Paul Boyer, the Boy Scout movement was an integral part of the
    Progressive era positive environmentalist effort. See Boyer, *Urban Masses,*
    359, no. 61.

69. Ernest Thompson Seton, *Boy Scouts of America: A Handbook of Woodcraft,
    Scouting, and Lifecraft* (New York, 1910), xi, xii, 1, 2; as quoted in Nash,
    Wilderness and the American Mind, 148.

70. I have found no historical literature pertaining to Franklin Roosevelt's experi-
    ences with the Boy Scouts. The quote is from Franklin Roosevelt to Conrad
    Chapman, June 15, 1925, FDR: Family, Business and Personal, Subject File:
    Boy Scout Federation of Greater New York, Correspondence: A–C, FDRL;
    other facts were pieced together from the following material: Colin
    Livingston to Franklin Roosevelt, May 16, 1921, FDR: Family, Business and
    Personal, Subject File: Boy Scout Federation of Greater New York,
    Correspondence: D–M, FDRL; Boy Scout Foundation of Greater New York,
    "Annual Report, January 1923," FDR: Family, Business and Personal, Subject
    File: Boy Scout Federation of Greater New York, Correspondence: N–W,
    FDRL; and "Roosevelt Quits Presidency of Boy Scout Unit," *New York
    Herald Tribune,* July 22, 1937, Clipping From Presidents Personal File,
    #4241, Boy Scout Federation of Greater New York, FDRL.

71. "Roosevelt Sees Problem of Boys Aided by Scouts," *New York City Evening
    World,* March 2, 1929, FDR: Papers as Governor of New York State, Series 1:
    Correspondence, Boy Scout Foundation of New York, FDRL.

72. Franklin Roosevelt, "Magnitude and accomplishment of the Boy Scout
    movement," address given over radio station WJZ at the Luncheon of the Boy
    Scouts Foundation in New York City, April 8, 1932, Speech File #471,
    FDRL.

73. Franklin Roosevelt, "How Boy Scout Work Aids Youth," *New York Times,*
    August 12, 1928, FDR: Family, Business and Personal, Subject File: Boy
    Scout Federation of Greater New York, Correspondence: Proctor, Arthur, W.,
    1925–1928, FDRL.

74. The Boy Scouts established the first of these campgrounds in 1917. The
    18 campsites functioning in 1922 could accommodate anywhere from 75 to

450 boys each. For statistics concerning the Boy Scout Foundation of Greater New York's campgrounds, see Boy Scout Foundation of Greater New York, "Annual Report, January 1923," FDR: Family, Business and Personal, Subject File: Boy Scout Federation of Greater New York, Correspondence: N–W, FDRL; A.C. Olson to Franklin Roosevelt, July 27, 1921, FDR: Family, Business and Personal, Subject File: Boy Scout Federation of Greater New York, Correspondence: N–W, FDRL; Franklin Roosevelt to James Forbes, August 1, 1921, FDR: Family, Business and Personal, Subject File: Boy Scout Federation of Greater New York, Correspondence: N–W, FDRL; and A. Schaeffer, Jr., "A Hotel that is as Large as All Outdoors," *National Hotel Review,* nd, FDR: Family, Business and Personal, Subject File: Boy Scout Federation of Greater New York, Correspondence: N–W, FDRL.

75. Franklin Roosevelt, "Magnitude and accomplishment of the Boy Scout movement," address given over radio station WJZ at the Luncheon of the Boy Scouts Foundation in New York City, April 8, 1932, Speech File #471, FDRL. According to historian David Macleod, during the 1910s and 1920s the Boy Scouts nationwide were plagued by a lack of campground sites. See, Macleod, *Building Character in the American Boy,* 242.

76. Franklin Roosevelt to James Forbes, August 1, 1921, FDR: Family, Business and Personal, Subject File: Boy Scout Federation of Greater New York, Correspondence: N–W, FDRL.

77. For a description of the Sullivan County Boy Scout campground, see "Roosevelt Sees Problem of Boys Aided by Scouts," *New York City Evening World,* March 2, 1929, np; and "Scouting Solves Juvenile Crime, Says Roosevelt," *Brooklyn New York Eagle,* March 2, 1929, np; both found in clipping file, FDR: Papers as Governor of New York State, Series 1: Correspondence, Boy Scouts of America, FDRL.

78. Boy Scouts of America, "Fifteen Million American Boys Call to You . . ." (1929), clipping found in FDR: Papers as Governor of New York State, Series 1: Correspondence, Boy Scouts of America, FDRL.

79. Macleod, *Building Character in the American Boy,* 239.

80. Ibid., 245. For a general description of the destructive character of Boy Scout camping, see Macleod, Ibid., 140 and 239.

81. Franklin Roosevelt to George Pratt, September 6, 1922, FDR: Family, Business and Personal, Subject File: Boy Scout Foundation of Greater New York, Correspondence: N–W, FDRL.

82. Roosevelt's request was mentioned in the following letter: Louis Howe to H.A. Gordon, July 13, 1922, FDR: Family, Business and Personal, Subject File: Boy Scout Foundation of Greater New York, Folder: FDR Conservation Camps, FDRL.

83. Director of Camp Museums, Palisades Interstate Park to H.A. Gordon, July 24, 1922, FDR: Family, Business and Personal, Subject File: Boy Scout Foundation of Greater New York, Proctor, Arthur, W., 1922–1924, FDRL.

84. For a good description of the Franklin D. Roosevelt Conservation Camps, see Chairman, Camp Committee, Boy Scout Foundation of Greater New York to Louis Howe, nd, FDR: Family, Business and Personal, Subject File: Boy Scout

Foundation of Greater New York, Folder: FDR Conservation Camps, FDRL. For promotional literature on the conservation camps, see "Scout Camps: F.D. Roosevelt Conservation Camps, Harriman Section, Palisades Interstate Park," pamphlet found in FDR: Family, Business and Personal, Subject File: Boy Scout Foundation of Greater New York, Folder: FDR Conservation Camps, FDRL.

85. Fay Welch to Edgar Nixon, April 18, 1955, FDR: Family, Business and Personal, Subject File: Boy Scout Foundation of Greater New York, Folder: FDR Conservation Camps, FDRL. On Roosevelt's desire to include a forestry program at the Boy Scouts Sullivan, County, New York campground, see Arthur Proctor to Guernsey Cross, August 26, 1930, FDR: Papers as Governor of New York, Series 1: Correspondence, Boy Scout Foundation of Greater New York, FDRL.

86. On the Dutchess County, New York project, see Walter Forse to Franklin Roosevelt, December 16, 1931, FDR: Papers as Governor of New York State, Series 1: Correspondence, Boy Scouts of America, FDRL. On the Boy Scouts nationwide "Nut Seed and Tree Planting Project," see "Boy Scouts are Undertaking Wide Tree-Planting Project," *New York Times,* April 20, 1930, sec. ix, 8; and "Boy Scouts Are Embarked on Tree-Planting Campaign," *New York Times,* June 1, 1930, sec. viii, 12.

87. Franklin Roosevelt to the New York State Legislature, March 25, 1930, as reprinted in Nixon, *FDR and Conservation,* 1: 71.

88. Governor Roosevelt established TERA as part of the Wicks Act of September of 1931. For background information on the Temporary Emergency Relief Administration (TERA), see Bellush, *Franklin D. Roosevelt as Governor of New York,* 141–49; Potter, "The Civilian Conservation Corps in New York State," 25; and John Gibbs, "Tree Planting Aids Unemployed," *American Forests* (April 1933): 159–61.

89. Franklin Roosevelt to Ovid Butler, August 15, 1932, as reprinted in Nixon, *FDR and Conservation,* 1: 122.

90. "New York Employs 10,000 Men for Tree Planting," *American Forests* (August 1932): 467; "State Speeds Jobs in Reforestation," *New York Times,* July 25, 1932, 17; and Gibbs, "Tree Planting Aids Unemployed," 159.

91. On TERA conservation work and the 25,000 unemployed men aided by the program, see Gibbs, "Tree Planting Aids Unemployed," 161 and 159. In 1932 TERA employed approximately 1,000 New York City men in the Palisades Interstate Park. See, "Thousand New York City Men Given Work in Interstate Park," press release from the Commissioners of the Palisades Interstate Park, 1932, Bear Mountain State Park Archives, Bear Mountain, New York.

92. On TERA work camps and weekly pay, see Potter, "The Civilian Conservation Corps in New York State," 28. For TERA's role as CCC selection agent, see Potter, "The Civilian Conservation Corps in New York State," 27. According to Potter TERA maintained ten camps in central New York for its workers.

93. President Hoover allocated a small amount of funds to unemployment relief work in National Forests in January of 1931, approximately one year *after* Governor Roosevelt asked the New York legislature for an appropriation for tree-planting relief work. On President Hoover's forestry relief efforts, see "Forestry to Help Unemployment," *American Forests* (January 1931): 48; and "Work on National Forests Contributes to Relief of Unemployment," *American Forests* (April 1931): 232.

94. F.A. Anderson, Executive Committee, Mississippi Forestry Association to Franklin Roosevelt, November 5, 1931, as reprinted in Nixon, *FDR and Conservation,* 1: 99.

95. For an overview of these states' forestry work-relief programs, see G.H. Collingwood, "Forestry Aids the Unemployed," *American Forests* (October 1932): 550. Other states that helped unemployed families by allowing them to collect cordwood and other wood products from state-owned land include New Hampshire, North Carolina, Connecticut, Louisiana, Utah, Idaho, and Colorado.

96. California Governor James Rolph established California's forestry work-relief program on November 27, 1931. For statistics on the California program for 1932, see Samuel Blythe, "Camps for Jobless Men," *Saturday Evening Post,* May 27, 1933, 9; Other primary sources on California's forestry work camps include: Winfield Scott, "California's Unemployment Forest Camps," *American Forests* (February 1933): 51–54; R.L. Deering, "Camps for the Unemployed in the Forests of California," *Journal of Forestry* 30 (1932): 554–57. Secondary sources on California's forestry work-relief program include: Raymond Clar, *California Government and Forestry,* 2: 202. The California program began on December 1, 1931; Loren Chan, "California During the Early 1930s: The Administration of Governor James Rolph, Jr., 1931–1934," *Southern California Quarterly* 63, no. 3 (1981): 272–76; Alison Otis, *The Forest Service and the Civilian Conservation Corps, 1933–1942* (Washington, DC: United States Government Publishing, need date), 5; Amelia Fry, *National Forests in California* (Berkeley: University of California, Berkeley, 1965), 4; Potter, "The Civilian Conservation Corps in New York State," 29.

97. On the Progressive era's link to the history of production, see Hays, *Beauty, Health, and Permanence,* 13.

98. Franklin Roosevelt, "Informal Extemporaneous Remarks of the President, Ten Mile River Camp of the Boy Scouts of America, Ten Mile River, New York," August 23, 1933, FDR: Speech File #647, FDRL.

PART 2

CONSERVATION: WILDERNESS, AGRICULTURE, AND THE HUMAN COMMUNITY

CHAPTER 4

NEW DEAL
CONSERVATION: A
VIEW FROM THE
WILDERNESS

PAUL SUTTER

IN LATE 1934 AND EARLY 1935, JUST AS MANY OF FRANKLIN D. ROOSEVELT'S (FDR) conservation initiatives were taking shape, several of the era's most distinguished environmental activists—including Aldo Leopold, Bob Marshall, Benton MacKaye, and Robert Sterling Yard—came together to form the Wilderness Society, the first national organization dedicated to the creation and protection of a system of wilderness areas on America's public lands. Over the previous decade and a half, the founders of the Wilderness Society had cobbled together what we know today as the modern wilderness idea, one that found a statutory home in the Wilderness Act of 1964, but the New Deal context was crucial to the coalescence of modern wilderness advocacy. The single most important facet of the Wilderness Society's early advocacy was the group's united opposition to the threats that roads and automobiles posed to the nation's remaining wildlands. These pioneer wilderness advocates were "driven wild," pushed into a new brand of preservationist advocacy by a growing love affair between Americans, their automobiles, and wild nature.[1] Such threats to wilderness were pronounced throughout the interwar era, but they climaxed during the early New Deal as the federal government deployed an army of unemployed workers on the public lands, many of them charged with building roads, trails, and other modern improvements designed to open the public domain to motorized outdoor recreation.

While organized wilderness advocacy was one of the signature environmental achievements of the New Deal era, it was also a development in tension with many of the conservation initiatives of the Roosevelt administration. This essay examines that tension between New Deal conservation and wilderness preservation, and suggests that wilderness advocacy embodied an important critique of the New Deal environmental program. But to see these wilderness advocates only as critics of the New Deal is to miss much of their larger value to the project of recovering the environmental legacy of FDR's New Deal. For all their concerns about wilderness, most of the founders of the Wilderness Society were avid supporters of the New Deal conservation agenda—at least in principle. Many worked in one capacity or another for New Deal conservation agencies, and most saw the Roosevelt administration as one filled with tremendous promise for protecting the environment and reforming human–environmental relationships. The founders of the Wilderness Society, then, were not only critical observers of the threats posed to wilderness by New Deal conservation; they were also important participants in and commentators on the broader contours of conservation policy and practice under FDR. Their vantage points—their views from the wilderness—can help to make sense of an era that environmental historians have thus far left largely unexplored.

Compared with the Progressive and postwar eras, the New Deal has been neglected by historians of environmental thought and politics. To this day, despite some excellent focused studies on various New Deal conservation agencies and efforts, we lack an adequate synthetic history of New Deal conservation.[2] Even those who have studied the substance and legacy of New Deal politics have spent little time on conservation, despite its centrality to the New Deal agenda. Most scholars, following the lead of Samuel Hays, have assumed that the rise of utilitarian conservation during the Progressive era and the emergence of a broad-based environmental movement after World War II were the two most important developments in environmental thought and politics during the twentieth century. Indeed, Hays posited a fairly smooth transition "from conservation to environment"—a transition that mirrored the broader historical shift from a producer to a consumer economy and culture—that has left little room for seeing New Deal conservation as anything more than an extension of traditional conservation concerns or a vaguely transitional period.[3] Scholars who have looked at interwar environmental thought and politics, in other words, have generally looked either at how Progressive conservation agendas fared during the era, or for early hints of postwar environmental sentiment. Rarely have scholars asked how New Deal conservation was unique, or how a more thorough understanding of its terrain might bring into relief the broader contours of twentieth century environmental history.

Research into the careers and thought of the founders of the Wilderness Society, however, suggests a whole series of ways of making sense of New Deal conservation. Aldo Leopold embodied much of what was new and important about the wildlife policies of the 1930s, for instance, and he was also a keen observer of the changes taking place in America's agricultural landscapes, of the dangers of trying to push the homestead ideal into environments that were agriculturally marginal, and of the need to restore those parts of the landscape that had been abused. Benton MacKaye, who worked for the Tennessee Valley Authority (TVA) for almost two years, was representative of some of the social and technical aspects of New Deal conservation. He, like many interwar conservationists, was attracted to both resettlement and regional planning, and he also expressed a technological optimism that was common currency during the interwar period, an optimism that, for a variety of reasons, was lost to postwar environmentalism. Bob Marshall spent the New Deal years working as a forester for the Bureau of Indian Affairs (BIA) as John Collier was remaking federal Indian policy, and he then became the first head of the Forest Service's newly created Division of Recreation and Lands in 1937. In both capacities, Marshall played an important role in shaping New Deal conservation policy, and in both areas the New Deal offered some important innovations. By looking through the eyes of these three activists, this author hopes to suggest some new ways of thinking about New Deal conservation and its legacy.

"This country has been swinging the hammer of development so long and so hard," Aldo Leopold wrote in the September 1935 issue of *The Living Wilderness*, "that it has forgotten the anvil of wilderness which gave value and significance to its labors. The momentum of our blows is so unprecedented that the remaining remnant of wilderness will be pounded into road-dust before we find out its values."[4] Like many of his colleagues who contributed to this first edition of the Wilderness Society's magazine, Leopold imbued his article with a sense of alarm and urgency, and his use of the "road-dust" image made clear that he, like the organization's other founders, was particularly concerned with how roads and automobiles had penetrated and carved away at the nation's remaining wildlands. Indeed, one cannot help but be struck, in reading this initial issue of their magazine, by the founders' overwhelming characterization of roads as the primary threat to wilderness.

A decade and a half earlier, Leopold, then a forester with the U.S. Forest Service in the Southwest, had been the first person to give voice to these concerns, and to suggest the creation of wilderness areas on the national forests as a protective measure against road building and motorized recreation. In the interim, he and like-minded activists had had some success convincing the Forest Service to pursue a policy of wilderness preservation.[5] But, for Leopold, the advent of the New Deal presented a new set of threats. "There is a particular need for a Society now," Leopold wrote later in his *Living Wilderness* piece, "because of the pressure of public spending for work relief.

Wilderness remnants are tempting fodder for those administrators who possess an infinite labor supply but a very finite ability to picture the real needs of his [*sic*] country."[6] While the underlying phenomena that had given wilderness preservation its meaning were several decades old by 1935, Leopold made clear that the New Deal's aggressive approach to conservation work—an approach that prioritized the mobilization of workers over thorough land planning—had precipitated the formation of the Wilderness Society.

Other contributions to the first *Living Wilderness* were equally clear on the threat of New Deal conservation measures to wilderness conditions. Robert Marshall published an open letter to Lithgow Osborne, New York State's Conservation Commissioner, protesting Osborne's proposal to utilize Civilian Conservation Corps (CCC) labor to construct an extensive network of "truck trails" throughout the Adirondack State Park, Marshall's beloved boyhood stomping grounds. Such roads, Marshall argued, would provide little tangible benefit in terms of fire protection—in fact, he suggested that the recreational access such roads would provide to motorists might even increase the fire hazard to the region—and they would do serious damage to the wilderness qualities of the park. Moreover, Marshall charged that the scheme was largely an effort to stake a claim to relief funds that might otherwise go elsewhere. In other words, Marshall intimated that New Deal conservation rested on a brand of pork-barrel politics that created incentives for make-work projects that often did not make conservation sense—and that undermined wilderness conditions in the process.[7]

Several other pieces in the first *Living Wilderness* focused on the New Deal affection for recreational parkways. Of particular concern were a series of skyline roads being planned for and built along the Appalachian ridgeline, roads that threatened the integrity and primitive qualities of a competing recreational plan for the region—the Appalachian Trail (AT). Wilderness Society founder Benton MacKaye had first proposed an Appalachian Trail in a 1921 article in the *Journal of the American Institute of Architects,* and by the early 1930s the trail was nearing completion (in fact, CCC workers would complete the final stretch in Maine in 1937, though the damage done by the Hurricane of 1938 was so severe that the trail would not be completely connected again until after World War II).[8] In "Why the Appalachian Trail," MacKaye urged the Trail's supporters to defend the AT and its wilderness environment against the depredations of scenic roads and the ever-widening "zone of motor influence." Indeed, MacKaye had come to understand the importance of wilderness preservation in the crucible of a clash between the proponents of motorized recreation, for whom the New Deal was a flush time, and hikers who sought, in his words, "a refuge from the crassitudes of civilization—whether visible, tangible, audible—whether of billboard, of pavement, of auto-horn."[9] Again, the motorization of outdoor recreation had preceded the New Deal era by several decades, but for MacKaye, like his colleagues, the federal government's willingness to build an infrastructure to support motorized outdoor recreation represented a new threat to wilderness, a threat that drove them to organize.

Harold Anderson, another of the founders of the Wilderness Society and a prominent member of the Potomac Appalachian Trail Club, echoed many of MacKaye's sentiments in his essay, "Primitive Trails and Super-Trails." Anderson's particular concern was with how the Skyline Drive in Shenandoah National Park—a project begun during the Hoover administration but completed during the New Deal—had thoroughly intruded upon the primitive qualities of the AT as it snaked through the park. Rarely was the trail more than several hundred yards from the road and its din, Anderson complained. As importantly, much of the trail had been widened and graded so that it no longer resembled the "primitive trail" that he and others had sought. In this objection, Anderson intimated that the landscaping of national parks and other public wildlands for recreational ease, a process that kept many a New Deal relief worker busy, undermined the ideal of wilderness recreation—one of rugged travel through a natural environment that had few improvements and was isolated from the signs and conditions of modernity. Shenandoah was just one of a number of places where the possibilities for wilderness recreation seemed to be disappearing because of recreational rather than resource development.[10]

In perhaps their biggest coup, the founders received permission from Secretary of the Interior Harold Ickes to reprint in the first issue of the *Living Wilderness* cautionary comments he had made about road building in the national parks. "I am not in favor of building any more roads in the National Parks than we have to build," Ickes maintained. Nor was Ickes "in favor of doing anything along the line of so-called improvements that we do not have to do." "This is an automobile age," he continued, "but I do not have a great deal of patience with people whose idea of enjoying nature is dashing along a hard road at fifty or sixty miles an hour." Ickes appreciated not only the concern that New Deal efforts to outfit the public lands for motorized recreation often came at the expense of wilderness conditions, but he also understood the critique of motorized recreation that lay behind it—that the automobile and scenic roads worked a qualitative transformation of the aesthetic experience of traveling through wild nature. Harold Ickes emerged as wilderness advocates' most important New Deal ally, and his sympathies with the Wilderness Society's program had much to do with keeping a skyline drive out of the Great Smoky Mountains National Park and with the establishment of several new national parks on a wilderness model.[11]

The founders of the Wilderness Society also sought to contain what Robert Sterling Yard saw as a dominant "administrative fashion" during the

New Deal: various attempts "to barber and manicure wild America as smartly as the modern girl."[12] If the proliferation of a network of roads and the automobiles that used them fostered in these advocates a commitment to saving large, roadless areas, the trend towards manicuring wild nature gave them another reason to elevate and celebrate an aesthetic of wildness. The New Deal was a good time to be a landscape architect, at least if one's goal was to get work with a federal conservation agency. Yard's particular concern was the national parks. Richard Sellars, in his history of the National Park Service, has estimated that the Park Service employed as many as 400 landscape architects during the height of the New Deal. Prior to the New Deal, the Park Service had employed only a handful of such landscape professionals; only nine had been on staff as late as 1929. But the rapid increase in the Park Service's overall workforce—which grew from a monthly average of about 2,000 to well over 17,000 between 1932 and 1936—demanded such design oversight.[13] These landscape architects, and the CCC workers they oversaw, created an extensive built landscape in the national parks. As importantly, landscape architects attempted to smooth the rough edges between motorized access and modern amenities on the one hand and wild nature on the other. This was necessary work if the parks were to absorb both development and a large numbers of visitors without losing their naturalistic feel, but the ubiquity of such work also brought the value of wilderness—as unmanicured wild nature—into sharp relief. The founders thus defined wilderness not only as roadless and unmotorized, but also in opposition to another signature New Deal feature: the omnipresence of landscape design in supposedly wild nature.

The founders of the Wilderness Society were not unconditionally opposed to scenic roads and motorized recreation; they appreciated the need to open up the public lands to a populace demanding opportunities for outdoor recreation. But the vast majority of the nation's most scenic public lands had already been so opened, and the founders were not willing to see the few large roadless remnants sacrificed to this craze. Protecting wilderness, then, involved careful planning. The founders of the Wilderness Society also recognized that leaving wild nature alone was a particular challenge at a moment when Americans' pent-up recreational energy collided with the New Deal's radical expansion of federal capacity to undertake conservation work, much of it directed at recreational development. During the 1920s, the federal government had flirted with a broader role in providing for recreation, most notably during the two meetings of the National Conference on Outdoor Recreation in 1924 and 1926, but the resources and labor power for such an effort only materialized during the New Deal. The founders thus argued for careful federal planning, and that federal planners needed to give wilderness a place in a developing federal (and state) recreational landscape that also provided for modern recreational needs and desires.

The federal government's increased capacity to undertake conservation work during the New Deal, and particularly recreational development, had much to do with the exigencies of the Depression and the political support for work relief that flowed from depression conditions. But it's also important to note that, even prior to the Depression and the New Deal, political support for preserving and developing national parks and recreational areas within national forests had been building. For much of the late nineteenth and early twentieth centuries, preservation had been a hard sell because it shielded land and resources from those who would develop them. And opposing economic development was not politically popular. But after World War I, politicians and civic leaders increasingly recognized that a national park or other federal recreational area in one's district not only brought federal dollars for developing roads and facilities, but also promised the profits that tourism would bring. The centrality of recreational development to New Deal conservation thus rested on a deeper political current: preservation became increasingly palpable as a political achievement that brought economic rewards.[14] Part of the function of the wilderness idea during the New Deal, then, was to challenge the commercial logic behind a new political interest in preservation and outdoor recreation.

Commerce, however, was not the only motivation behind New Deal efforts to develop the public lands for recreation. Outdoor recreation had itself come to have important cultural meaning during the early twentieth century, and one of the achievements of the Roosevelt administration was its successful incorporation of that meaning into the politics of national recovery. For most of the nation's history, from Thomas Jefferson to Frederick Jackson Turner, work in nature had been celebrated as virtuous, but as industrialization degraded work, as urbanization moved more Americans away from a direct working relationship with nature, and as the frontier waned, Americans increasingly invested leisure in nature with the sorts of virtues that work in nature had once embodied. Figures such as Theodore Roosevelt had celebrated this ideal of recreational strenuosity (and masculinity), but it was his distant cousin Franklin who made it a centerpiece of his conservation policy—not only by developing more opportunities for such recreation, but also by making opportunities for paid working vacations in nature a central part of his work-relief offerings. New Deal conservation, in other words, was animated by a complex alchemy of work, leisure, and nature at a moment when their cultural meanings were in flux.

Out of the objections by wilderness advocates to New Deal conservation projects and policies emerge several important points about New Deal conservation. First, let me reiterate that one of the great achievements of the New Deal, as Phoebe Cutler and others have shown, was the creation of a vast new public landscape, much of it on the nation's wildlands and in the service of outdoor recreation. Agencies such as the CCC and the Works Progress Administration provided unprecedented labor and funding for road building and recreational development within the national parks and national forests, for the protection and development of new landholdings such as national seashores, and for dramatically expanding state park systems (a topic that itself cries out for more historical attention). And all of this activity reflected, served, and perhaps even shaped the era's dramatic expansion of interest in outdoor recreation, an interest whose realization was facilitated by the automobile's proliferation. Federal programs during the New Deal were thus crucial to the "construction" of new recreational relationships between Americans and the natural world, relationships that simultaneously increased political support for preservationist programs and reframed nature as a tourist or consumer destination.[15] This new public recreational landscape and the cultural and political changes that underlay it were among the important and lasting legacies of the New Deal.

It is perhaps not surprising that wilderness advocacy should have emerged out of this particular historical moment, when Americans were increasingly enthusiastic about outdoor recreation and supportive of the dedication and development of large portions of the public domain to recreational uses. What is surprising, however, is that wilderness advocacy arose as a corrective to these trends. In the simplest sense, the New Deal emphasis on recreational preservation and development can be seen as part and parcel of that shift from conservation to environment described by Samuel Hays—from a federal conservation policy dominated by producerist and expert-driven notions of wise resource use to one increasingly focused on recreational provisioning of the sort desired by a nation of consumers. But the organized wilderness advocacy of the 1930s stretches this model, for wilderness advocacy arose not so much in league with as in opposition to recreational consumerism. The founders of the Wilderness Society certainly were concerned with the impacts of resource exploitation on the nation's wildlands, but that concern had little to do with the conceptual innovation at the core of the wilderness idea: that certain wildlands ought to be protected from road building, motorized recreation, and a state increasingly eager to sponsor outdoor recreational development. The era's most important preservationist development was essentially a critique of the very trends that most scholars have assumed drove the growth and evolution of preservationist politics. Or, to put it more broadly, we need to understand the preservationist politics of the New Deal not simply as an accommodation to ascendant consumer preferences, but as a struggle to make sense of what it would mean to preserve nature in the context of such growing interest in outdoor recreation.

Seeing wilderness advocacy as a critical response to consumer trends in outdoor recreation suggests another important point: distinctions and tensions internal to the politics of outdoor recreation (and, for that matter, those internal to resource conservation) may well be more salient in making sense of New Deal environmental thought and politics than conflicts between the politics of preservation and use. Progressive conservation climaxed in a conflict over whether to build a dam in the Hetch Hetchy Valley within the bounds of Yosemite National Park. It was a battle that pitted arch-preservationist John Muir against the chief ideologue of wise resource use, Gifford Pinchot, and it seemed to set in stone the orthodoxy that the major ideological division within the environmental community has been between those who would preserve nature as it is and those who would wisely develop its resources. That the first major conservation battle after World War II—the fight over the Bureau of Reclamation's efforts to put a dam in Echo Park, within the boundaries of Dinosaur National Monument—seemed to mirror the Hetch Hetchy episode in all but outcome has only reinforced (and perhaps helped to produce) the traditional historiographical focus on the political clashes between preservation and use. Not only that, but the similarity between these two episodes has functioned as a scholarly bypass around the interwar years. During the New Deal, conflicts between preservation and utilitarian conservation were not nearly as salient. In fact, one of the most intriguing aspects of the era's wilderness advocates was that quite a few of them were trained and committed foresters and among the era's most important thinkers on the problems of human land and resource use. They did not see any ideological inconsistencies in being both preservationists and conservationists. Many New Deal conservationists felt similarly. The conservation–preservation dichotomy broke down during the 1930s, and closer attention to New Deal conservation may well reveal that we have overdrawn the extent to which such a dichotomy defined Progressive and postwar environmental thought and politics.

Now that we have examined how wilderness advocacy both highlighted and critiqued some of the signature developments of interwar conservation, it is time to turn to other ways in which the careers and thought of these advocates suggest ways of understanding New Deal conservation and its legacy. Here, three figures in particular are especially important: Leopold, MacKaye, and Marshall—each of whom refused to be typecast as either a conservationist or a preservationist.

Aldo Leopold is perhaps best known for his posthumously published *A Sand County Almanac,* arguably the most important piece of American nature writing of the twentieth century. One of the notable achievements of that work is its narration of an ethical shift away from anthropocentrism and toward biocentrism, a shift informed by the science of ecology. In historical terms, that shift manifests itself most notably in a couple of areas of federal conservation policy during the New Deal. First, and perhaps most famously, it was reflected in wildlife policy, and in a transition from a notion of "game" to one of "wildlife" management—a shift that challenged the efficacy of distinctions between species that were and were not useful to humans, and laid the groundwork for reconsidering predator control policies. Game or wildlife management—a development to which Leopold, the nation's first professor of game management and the author of an authoritative text of the subject, was a crucial contributor—was central to the missions of New Deal agencies such as the CCC and the Soil Conservation Service (SCS). Moreover, several important pieces of wildlife legislation became law during the 1930s, significant acreage was added to the nation's system of wildlife refuges, and by 1940 the work of the Bureau of Fisheries and Biological Survey had been moved to Interior and combined to produce the Fish and Wildlife Service. Indeed, as Thomas Dunlap has suggested, the Roosevelt administration "made wildlife preservation an integral part of conservation programs" in unprecedented ways.[16] Wildlife protection and habitat development ought to be seen as another signature conservation development during the New Deal, as well as a variable that complicated federal environmental management. Indeed, Leopold was vocal about New Deal conservation initiatives that failed to take into account how the needs of wildlife demanded management techniques not always consonant with other conservation goals, such as silvaculture.

Aldo Leopold's career and thought also suggest how central the nation's agricultural landscapes and farming practices were to New Deal conservation. In the most basic sense, New Deal conservation was innovative in its efforts to move beyond the public lands to push and even direct conservation initiatives in the private agricultural landscape. And as Donald Worster has shown, New Deal agricultural conservation involved the shotgun marriage of ecology and technocratic managerialism. Scientists such as Paul Sears popularized an ecological critique of agricultural expansion and mechanization, while technocrats such as Hugh Hammond Bennett made soil conservation a central part of Roosevelt's conservation agenda.[17] Both of these subjects, particularly the history of soil conservation during and after the New Deal, deserve more scholarly attention.[18]

But there is a broader point about agricultural conservation during the New Deal that is necessary to consider: New Deal conservationists found themselves saddled with the responsibility of responding to the ecological and social results of several decades of agricultural expansion into marginal environments. This process was most obvious on the southern plains, a region into which Americans had flooded during the early twentieth century, served by generous federal land-disposal policies, armed with new agricultural technologies, and buoyed by high grain prices. The results were disastrous, compelling the federal government to respond with both relief and technical assistance that (along with the Ogallala Aquifer) allowed much of the dust bowl region to be kept in crops. But the worst lands reverted to public management and became the seeds for a new system of national grasslands. This process was occurring elsewhere as well—in the Great Lakes states, for instance, where farmers had moved into the cutover and attempted to make such lands agriculturally productive. Many of those efforts failed, a fact that Aldo Leopold knew only too well. In 1935, he and his family purchased some land in Sauk County, Wisconsin that had been cut over, farmed briefly and unsuccessfully, and then abandoned. Leopold's meditations on restoring that piece of land make up the bulk of *A Sand County Almanac*. And while Leopold kept his particular patch of earth in private ownership, much of the surrounding country reverted (often through abandonment or failure to pay taxes) to public ownership and management. To these examples of growing public responsibility for abused lands could be added many others—including areas that became part of the conservation landscape during the New Deal less as a result of agricultural expansion into marginal environments than as a result of a deeper history of land use and abuse. The Oconee National Forest in Georgia, for instance, has its roots in federal submarginal land purchases during the New Deal, as do other eastern national forests and many of the nation's state parks. The New Deal, in other words, was responsible for a significant expansion in the nation's publicly owned conservation landscape. Moreover, while most of those lands today serve traditional conservationist or preservationist ends, their early history was often one of *restoration*. Indeed, environmental restoration was another signature feature of New Deal conservation efforts. A more thorough appreciation of New Deal conservation would force scholars to recognize that restoration has been as important an environmental ideal as conservation and preservation, one that we can no longer afford to neglect.

The New Deal was also the point at which the federal government finally put an end to virtually all homesteading on the public domain. The Taylor Grazing Act of 1934 withdrew most of the nation's remaining public lands and placed them permanently in public ownership and under federal management—initially under the control of the Grazing Office, which gave way in 1946 to the Bureau of Land Management. The New Deal was thus an era in which federal environmental managers fully recognized the necessary permanence of a vast public domain that was marginal for traditional agriculture. To my mind, this suggests a way of rethinking the entire history of American conservation policy. From the first revisions of homestead policy through the Taylor Grazing Act, American conservation has been at least in part a process of grappling with environmental limitations posed by those portions of the landscape that were not amenable to traditional agricultural settlement patterns and land-disposal policies. Or, to put it another way, what Frederick Jackson Turner saw as the end of the frontier in 1890 might more accurately be seen as the end of arability, with conservation stepping in to address the problem in a variety of ways. Scholars have long seen early efforts to preserve wild nature and conserve natural resources as laudable correctives to the industrial exploitation of nature, or alternately as efforts aimed at elevating resource efficiency or providing for the recreational needs of consumers. But we might also see them as efforts to contend with an impending reality that much of the landscape—and particularly the arid and mountainous West—would remain, or in fact become, a public charge. The New Deal was the era in which this aspect of conservation became clear.[19]

While Aldo Leopold was often critical of New Deal conservation, such sentiments spoke as much to the new complexities of environmental management during the era—the competing agendas of various conservation initiatives and interconnections between public and private lands—as they did to the limitations of New Deal conservation practice. Indeed, the upshot for Leopold was that government conservation, however necessary, was itself a limited and limiting solution; the land ethic he urged upon Americans was about recognizing the need for individual responsibility rather than govern-mental surrogacy. That topic itself, which Donald Worster examined in *Dust Bowl,* deserves a broader application.

If Aldo Leopold was the most thoughtful and eloquent of the era's environmental thinkers, Benton MacKaye was perhaps the most unorthodox. MacKaye's ideas often strained the limits of the possible, but his intellectual and professional evolution speaks volumes about the differences between Progressive and New Deal conservation. Indeed, many of MacKaye's efforts to stretch the bounds of conservation policy and practice foreshadowed New Deal programs and techniques, though mostly MacKaye would be disappointed with the New Deal iterations of concepts with which he had long toyed.

Like Leopold, MacKaye earned a graduate degree in forestry and was among the nation's first generation of professional foresters. Unlike Leopold, however, he soon began focusing on what he called the social aspects of forestry, and particularly the need to square sustainable forestry practices with sustainable communities and labor equity. In the mid-1910s, MacKaye left the Forest Service to work for the Department of Labor, where he prepared a report that argued for colonizing portions of the public lands with coopera-tive resource communities. The report—published in 1919 as *Employment and Natural Resources*—had a couple of important premises.[20] First, MacKaye based his desire for such communities on a realization that the individualistic model of homestead settlement had been a flawed one, and would be doubly dangerous as homesteaders moved onto what remained of the public domain. In this fear he proved prescient. MacKaye insisted that cooperative settlement of such lands, with copious government assistance, was crucial to the lasting success of the settlement process. In other words, MacKaye argued for a much more active government role in stabilizing the relationship between Americans and the land—an argument central to New Deal conservation. Second, MacKaye insisted that Americans see a relationship between resource exploitation and human degradation. Conservationists could not hope to reform the use of land and resources, MacKaye insisted, without paying much greater attention to the human side of the story. In this contention, MacKaye presaged another important aspect of New Deal conservation idealism: that human exploitation and environmental degradation were part of the same process and had to be attacked as such, and that government could be a crucial actor in addressing both ills. Or, as New Dealer Arthur Raper put it in his study of agriculture and New Deal intervention in Greene County, Georgia: "Just as there is a human side to soil conservation, there is a soil side to the rehabilitation of farmers."[21] MacKaye was making similar points two decades earlier, not only about agriculture but also about the timber and mining industries, in arguing that the federal government ought to pursue resettlement efforts aimed at righting human–environmental relationships.

MacKaye's radical ideas about using the federal government to reform the relationship between land settlement, labor, and resource exploitation never found the legislative backing he hoped they would, and in the wake of the Red Scare following World War I, he found it necessary to repackage them. Luckily for MacKaye, he fell in among a crowd of architects and urban planners, such as Louis Mumford and Clarence Stein, who, together with MacKaye, formed the core of a group that went on to found the Regional Planning Association of America (RPAA) in 1923. Mumford, Stein, and others helped MacKaye to frame his 1921 article proposing the construction of an AT, an article that proposed not only a long recreational footpath connecting New England to the Southern Appalachians but also plans for a series of socialist recreational and resource communities along the route.[22] MacKaye's AT plan, in other words, built upon his earlier work and was much more than a recreational plan for the region. But before long, MacKaye found himself defending the AT against the incursion of modern roads and motor tourists, and he did so by embracing, and in fact doing much to define the tenets and methodologies of regional planning.

MacKaye saw regional planning as a new vessel for containing his old ideas, but he also used regional planning to significantly expand the scale of his thinking. Two aspects of MacKaye's regional planning thought, which he outlined most fully in his 1928 book *The New Exploration,* seem particularly worth noting in the context of giving definition to New Deal conservation.[23] First, MacKaye realized that the integrity of the conservation landscape— that portion of the natural world preserved in public ownership and devoted either to sustainable resource development or recreation—relied upon a sweeping command of the forces shaping entire regions. As roads and automobiles intruded upon the AT, MacKaye increasingly talked about the AT as a "wilderness" trail—a term that he had not used to describe the trail in his 1921 article proposing it. He became a wilderness advocate in response to these threats. But MacKaye also began talking about wilderness as part of a landscape continuum that also included the urban and the rural, a continuum that had to be planned in opposition to the spread of "metropolitanism"— a rough synonym for what we would today call sprawl. MacKaye's vision of the ideal planned landscape thus came to include not only wilderness areas, "wilderness ways," and resource communities, but also garden cities, townless highways, open space dams and levees to control sprawl, and compact regional cities. For MacKaye, one of the premises of regional planning was that all portions of the landscape had to be planned together and in a regional context that substituted natural for political boundaries. And unlike his urban-focused RPAA colleagues, MacKaye celebrated the wilderness ridgeline as the best perspective from which to view the region and regional processes. Such regional planning thought was a defining component of New Deal conservation, particularly in such places as the Tennessee and Columbia River valleys. More broadly, regional planning highlights what has been one of the most confounding features of New Deal conservation: its borderlessness. New Dealers often integrated traditional conservation policy with agricultural interventions, resettlement efforts, town planning, and other urban and human-centered initiatives. In other words, they, like MacKaye, often refused to see conservation as isolated and separate from other problems. Indeed, as Robert Dorman has intimated, New Deal conservation must be understood in relation to a vital regionalist movement that thrived during the interwar years.[24]

There was one other aspect of Benton MacKaye's regional planning thought that helps to illuminate New Deal conservation's particular profile: MacKaye, like many other planners of the interwar era, had a particularly sanguine view of a suite of technologies that he, following the Scottish botanist and town planner Patrick Geddes, referred to as "neotechnic" technologies—among them new power sources such as electricity, new alloys such as aluminum, and new transportation technologies such as (believe it or not) the automobile. Although MacKaye and his fellow regional planners insisted that none of these technologies could achieve their full potential without careful planning, and that their unplanned deployment might create more problems than they solved, he nonetheless hoped that they would together allow for a healthier and more decentralized pattern of settlement and a better standard of living for those living in rural America. MacKaye was simultaneously concerned that America's industrial cities were growing beyond human scale while its rural regions remained technologically backward in ways that promoted both poverty and environmental degradation. This is all to say that MacKaye, like many other environmental thinkers during the 1920s and 1930s, was animated by a technological optimism that faded from American environmental thought after World War II. Nowhere was this technological optimism more apparent than in the support MacKaye and those of his ilk showed for public hydropower development—what some referred to as "Giant Power"—in regions such as the Tennessee Valley. New Dealers generally saw such grand technological achievements as consonant with a more environmentally sensitive approach to regional development, and MacKaye was particularly taken by the multipurpose nature of dams in the context of river basin development—their ability to deliver inexpensive public power, to control flooding, to subsidize fertilizer production, and to decentralize industrial production. Such technological optimism, though it strikes many today as naïve and misguided, was central to New Deal conservation thought. The lines between nature and artifice were less tightly drawn during the era.[25]

Benton MacKaye greeted the arrival of the New Deal with tremendous enthusiasm. He initially attempted to secure a position with the CCC, largely because he hoped the program might fulfill his resettlement ideas of the late 1910s. When such a position failed to materialize, he took a job with the Bureau of Indian Affairs for a short while (where he met Aldo Leopold, who was also doing some BIA work) before securing a position with the agency that most excited him—the TVA. MacKaye spent a little less than two years with the TVA. MacKaye hoped that the TVA would do regional planning on a large and ambitious scale, but by 1936 he had grown disillusioned with its focus on facilitating economic growth at the expense of careful planning for the public good.[26] Nonetheless, MacKaye's brief experience with the TVA and his broader career as a conservationist and environmental planner suggest that one cannot contend with New Deal conservation and its legacy without assessing the role of regional planning in shaping the era's ambitious and sweeping environmental programs. The scale, scope, and content of the planning impulse behind New Deal conservation distinguish it from what came both before and after.

Like Leopold and MacKaye, Bob Marshall was a trained forester, though he took his training in the early 1920s, a more conservative time for the profession. While Marshall's greatest reputation was made as a wilderness advocate—during the 1930s he virtually single-handedly reformed and substantially expanded the Forest Service's system of wilderness areas— he was also one of the most important voices in opposition to the conservative drift of the forestry profession and of federal forestry. The 1920s had seen the strong Progressive era regulatory rhetoric fade, to be replaced by a cooperative approach to the timber industry. Yet from the late 1920s through his death in 1939 at the age of 38, Marshall defended what he saw as the profession's radical pedigree, and he particularly pushed for a vast expansion of the nation's publicly owned forestlands. While his views on the subject, offered most clearly in his 1933 book *The People's Forests,* were extreme, he nonetheless was an important voice in the debate over New Deal forestry policy.

Marshall's forestry career and thought during the 1930s suggest a number of lessons germane to the broader New Deal landscape. First, Marshall's radicalism, which he shared with a cadre of contemporary foresters, intimates that the soul of the profession was more contested than other scholars have suggested, and that more work needs to be done on connecting New Deal conservation to the intellectual radicalism of the era.[27] Second, Marshall was perhaps the most important voice arguing for more careful attention on the part of the Forest Service to recreational planning, a responsibility the agency had long neglected but finally took seriously during the New Deal. Indeed, in 1937 the Forest Service hired Marshall as its head of the newly created Division of Recreation and Lands. Third, Marshall was a perfect embodiment of one of my central points about the era: that resource conservation and preservation were not mutually exclusive or conflicting ideologies. But it was Marshall's work as head of the BIA's forestry division from 1933 to 1937 that provides the most intriguing lens for getting a broader view of New Deal conservation.

Bob Marshall joined the BIA just as John Collier was remaking federal Indian policy. The Indian Reorganization Act, one of the most important legislative achievements of the era, reversed half a century of federal policy, rooted in the Dawes Severalty Act of 1887, aimed at assimilating Native Americans and consigning them to family homesteads. Instead, Collier and the BIA emphasized both the restoration of tribal lands, communally owned, and the recognition of tribal culture and identity. As the BIA's head forester, Marshall's job was to help Native American groups to protect and develop resources in ways consistent with a number of defining aspects of New Deal conservation. His job mixed complicated social and environmental reforms among groups of people often living in poverty. He dealt with a domain recently redefined as permanently public. Native Americans had to contend with making a living from marginal lands, a situation that drew in federal environmental managers. Marshall also romanticized the communalism of tribal landownership and the types of conservation management it seemed to make possible. Indian lands were a good place, then, to watch many of the unique aspects of New Deal conservation in action.

Contending with New Deal conservation efforts on Indian lands will be crucial to giving broad synthetic definition to the environmental history of the era. Those who have looked at the subject—in particular the work by Richard White and Marsha Weisiger on grazing reforms among the

Navajos—have shown, among other things, how the clash between Native American land-use practices and the environmental ideals of federal land managers provide an important nexus for examining the cultural biases built into technocratic perceptions of environmental marginality and narratives of environmental degradation—perceptions and narratives that were a central part of the era and its subsequent historiography. If we are to see environmental marginality as a central part of the New Deal conservation story, we also need to appreciate that marginality is as much a cultural as an environmental concept, and a close reading of the Native American experience with conservation will help to make that clear. The New Deal was also an era in which regionalist thinkers idealized Native American groups as indigenous folk cultures rooted in place and particular relationships with the land. Finally, examining the Native American experience with state conservation will allow historians to test out interpretations about the colonial nature of the rise of the conservation state and its impacts upon people at the margins, interpretations that have been central to an emerging revisionist historiography of Progressive conservation and that have been important to non-U.S. environmental historiography as well.[28]

The founders of the Wilderness Society have much to teach us about the contours of New Deal conservation. In crafting and organizing to protect a new preservationist ideal, they argued that New Deal conservation threatened wilderness in a number of important ways and in so doing they presented a broad critique of society drifting toward a consumer relationship with nature and a government increasingly eager to cater to that emerging relationship. They questioned the need to develop recreational nature and to open it up to modern forms of access; they expressed concern about the ubiquity of landscape design as a technique for masking the incongruities involved in both preserving and developing wild nature; and they challenged the growing commercialism that animated the politics of outdoor recreation during the interwar era. As a critique of recreational trends, modern wilderness advocacy suggests both the centrality and the complexity of outdoor recreation to New Deal conservation.

The careers and thought of the founders also revealed how important environmental marginality was to New Deal efforts, and how often conservation during the era involved restoring nature rather than simply saving or conserving it. Aldo Leopold's career during the era highlights the rise of wildlife concerns, the importance of agricultural conservation, and the need to think about the achievements of New Deal environmental management in the context of a nation that was growing away from the land and a direct ethical relationship with it. Benton MacKaye's career and thought underline how intermixed social and environmental concerns were during the New Deal, how regional planning helped to make sense of those connections on a large scale, and how technology was an important part of New Deal environmental planning. Finally, Bob Marshall's example suggests the need to break down the ideological division between conservation and preservation, to examine the relationship between conservation and other brands of radical thought and politics during the New Deal, and to take a critical look at what the Native American experience with the conservation state during the era might tell us about the dissonances between federal conservation expertise and idealism on the one hand and a variety of local land-use regimes on the other. For all three of these wilderness advocates, the view from the wilderness was an expansive one.

NOTES

1. Paul S. Sutter, *Driven Wild: How the Fight Against Automobiles Launched the Modern Wilderness Movement* (Seattle: University of Washington Press, 2002).
2. While it provides an adequate topical overview of New Deal conservation, A.L. Reisch Owen's *Conservation Under F.D.R.* (New York: Praeger Publishing, 1983) does not provide the interpretive rigor the era deserves. Important case studies range from Donald Worster's classic, *Dust Bowl: The Southern Plains in the 1930s* (New York: Oxford University Press, 1979) to Karen Merrill's recently published *Public Lands and Political Meaning: Ranchers, the Government, and the Property between Them* (Berkeley: University of California Press, 2002). Several recent dissertations will help to fill in the picture. See Cornelius M. Maher, "Planting More than Trees: The Civilian Conservation Corps and the Roots of the American Environmental Movement, 1929–1942" (Ph.D. diss., New York University, 2001); Sarah Phillips, *This Land, This Nation: Conservation, Rural America, and the New Deal* (Cambridge University Press, forthcoming); and Marsha Weisiger, "Dine Bikeyah: Environment, Cultural Identity, and Gender in Navajo Country" (Ph.D. diss., University of Wisconsin, 2000).
3. See Samuel Hays, *Conservation and the Gospel of Efficiency: The Progressive Conservation Movement, 1890–1920* (Cambridge: Harvard University Press, 1959); Hays, *Beauty, Health, and Permanence: Environmental Politics in the United States, 1955–1985* (New York: Cambridge University Press, 1987); Hays, "From Conservation to Environment: Environmental Politics in the United States Since World War II," in Hal Rothman and Char Miller, eds., *Out of the Woods: Essays in Environmental History* (Pittsburgh: University of Pittsburgh Press, 1997).
4. Aldo Leopold, "Why the Wilderness Society?" *The Living Wilderness* 1, 1 (September 1935): 6.
5. Leopold first laid out his wilderness plan in his article, "The Wilderness and Its Place in Forest Recreational Policy," *Journal of Forestry* 19, 7 (November 1921): 718–21. On Leopold's early wilderness thought, see Paul S. Sutter, " 'A Blank Spot on the Map': Aldo Leopold, Wilderness, and U.S. Forest Service Recreational Policy, 1909–1924," *Western Historical Quarterly* 29, 2 (Summer 1998): 187–214.

6. Leopold, "Why the Wilderness Society?" *The Living Wilderness* 1, 1 (September 1935).

7. Robert Marshall, "Fallacies in Osborne's Position: An Open Letter to the Conservation Commissioner of New York," *The Living Wilderness* 1, 1 (September 1935): 4–5. In that same issue, Raymond H. Torrey, secretary of the Association for the Protection of the Adirondacks, penned a similar critique. See "Truck Trails in the Adirondacks," 3, 5.

8. Benton MacKaye, "An Appalachian Trail: A Project in Regional Planning," *Journal of the American Institute of Architects* 9 (October 1921): 325–30.

9. MacKaye, "Why the Appalachian Trail?" *The Living Wilderness* 1, 1 (September 1935): 7–8.

10. Harold Anderson, "Primitive Trails and Super-Trails," *The Living Wilderness* 1, 1 (September 1935): 8.

11. Harold P. Ickes, "Wilderness and Skyline Drives," *The Living Wilderness* 1, 1 (September 1935): 12.

12. "A Summons to Save the Wilderness," *The Living Wilderness* 1, 1 (September 1935): 1. This particular article has no byline, but the author is fairly certain that it was written by Robert Sterling Yard and has assumed so in the text of the essay. The concerns raised in the piece are consistent with Yard's concerns, and as editor of Wilderness Society publications, Yard was the logical author of such unsigned pieces.

13. Richard West Sellars, *Preserving Nature in the National Parks: A History* (New Haven: Yale University Press, 1997), 52, 141–42. Part of the increase in personnel was attributable to the Park Service assuming control over the Capital and Military parks, but most of it had to do with the influx of CCC funding and labor.

14. Many scholars of early national park history have pointed out that such commercial logic always existed. The author agrees, but also thinks the New Deal era was crucial for tipping the balance in this direction in terms of broad political support for park preservation.

15. A number of recent studies have recognized the importance of the New Deal era's production of a built environment for a brand of outdoor recreation centered on automobility. See, e.g., Ethan Carr, *Wilderness By Design: Landscape Architecture and the National Park Service* (Lincoln: University of Nebraska Press, 1999); Linda Flint McClelland, *Building the National Parks: Historic Landscape Design and Construction* (Baltimore: Johns Hopkins University Press, 1997).

16. Thomas Dunlap, *Saving America's Wildlife: Ecology and the American Mind, 1850–1990* (Princeton, NJ: Princeton University Press, 1988), 82. See also Susan Flader, *Thinking Like a Mountain: Aldo Leopold and the Evolution of an Ecological Attitude toward Deer, Wolves and Forests* (Madison: University of Wisconsin Press, 1974).

17. See Worster, *Dust Bowl*.

18. See Sarah T. Phillips, "Lessons from the Dust Bowl: Dryland Agriculture and Soil Erosion in the United States and South Africa, 1900–1950," *Environmental History* 4, 2 (April 1999): 245–66; Neil Maher, " 'Crazy Quilt

Farming on Round Land': The Great Depression, the Soil Conservation Service, and the Politics of Landscape Change on the Great Plains during the New Deal," *Western Historical Quarterly* 31, 3 (2000): 319–39.

19. The author has discussed this issue at some length in two other essays: "Environmental History," in Stephen Whitfield, ed., *A Companion to Twentieth Century America* (London: Blackwell, 2004); "What Can U.S. Environmental Historian Learn from Non-U.S. Environmental Historiography," *Environmental History* 8, 1 (January 2003).

20. Benton MacKaye, *Employment and Natural Resources: Possibilities of Making New Opportunities for Employment through the Settlement and Development of Agricultural and Forest Lands and Other Resources* (Department of Labor, Washington, DC: GPO, 1919).

21. Arthur Raper, *Tenants of the Almighty* (New York: The MacMillan Company, 1943), 232.

22. MacKaye, "An Appalachian Trail," *Journal of the American Institute of Architects* 9 (October 1921).

23. MacKaye, *The New Exploration: A Philosophy of Regional Planning* (New York: Harcourt, Brace, and Company, 1928).

24. Robert Dorman, *Revolt of the Provinces: The Regionalist Movement in America, 1920–1945* (Chapel Hill: University of North Carolina Press, 1993).

25. Richard White, *The Organic Machine: The Remaking of the Columbia River* (New York: Hill and Wang, 1995).

26. Daniel Schaffer, "Benton MacKaye: The TVA Years," *Planning Perspectives* 5 (1990): 5–21.

27. Richard Pells, *Radical Visions and American Dreams: Culture and Thought in the Depression Years* (Urbana: University of Illinois Press, 1998 [1973]).

28. On the new Progressive era historiography, see Karl Jacoby, *Crimes Against Nature: Squatters, Poachers, Thieves, and the Hidden History of American Conservation* (Berkeley: University of California Press, 2001); Louis Warren, *The Hunter's Game: Poachers and Conservationists in Twentieth-Century America* (New Haven: Yale University Press, 1999); and Richard Judd, *Common Lands, Common People: The Origins of Conservation in Northern New England* (Cambridge, MA: Harvard University Press, 2000). On the relationship to non-U.S. historiography, see my essay, "What U.S. Environmental Historians Can Learn from Non-U.S. Environmental Historiography," *Environmental History* 8, 1 (January 2003): 109–29.

CHAPTER 5

FDR, Hoover, and the New Rural Conservation, 1920–1932

SARAH PHILLIPS

ON A JUNE AFTERNOON IN THE SUMMER OF 1931, FRANKLIN D. ROOSEVELT (FDR) ADDRESSED his fellow governors assembled in Indiana for their annual conference. The severe economic situation, he said, called for positive leadership, tangible experiments, and government guidance. Yet Roosevelt confined his remarks that day to one particular aspect of the nation's troubles—the "dislocation of a proper balance between urban and rural life." Describing how hundreds of farmers clung to exhausted lands and eked out an existence far below the "American standard of living," he outlined measures that New York State had initiated to classify lands, relieve tax burdens, purchase and reforest submarginal farmland, and bring cheaper electricity to the agricultural areas. He looked forward to a time when farmers cultivating land too worn to yield a profit would find alternate employment in factories close to rural communities. Planning for "a permanent agriculture," Roosevelt explained, was the state's ultimate purpose.[1]

Roosevelt's calls for "a proper balance" and "a permanent agriculture" anticipated the rural conservation programs of the New Deal. Conservationists in the Progressive era had essentially confined their efforts to waterways, forests, and recreational lands, and they advocated public ownership (mostly in the sparsely settled West) as the primary remedy for individual and corporate misuse. What made the New Deal different from these earlier efforts was its concern with inhabited rural regions and privately held land. The forces of unabated individualism, many policy makers concluded, had yielded an overinvestment in agriculture but an underinvestment in agriculturalists. The market had failed: farmers continued to till eroded and exhausted land, leaving the nation pockmarked with chronic rural poverty. "The throwing out of balance of the resources of Nature throws out of balance also the lives of men," Roosevelt declared. "We find millions of our citizens stranded in village and on farm because Nature cannot support them in the livelihood they had sought to gain through her."[2] The New Dealers therefore set out to rebuild rural life with measures tied directly to conservation objectives—land retirement, soil and forest restoration, flood control, and cheap hydropower for farms and small industries. Resource policy now had a dual purpose: the restoration of the land and the rehabilitation of its inhabitants.[3]

But these ideas did not spring forth, fully formed, in 1932. Roosevelt and his advisers drew from a new tradition of conservationist thinking, one largely developed in the decade before the New Deal. During the 1920s, two loosely connected groups of engineers, politicians, academics, and government officials voiced dissatisfaction with the nation's failure to meet the social and environmental needs of inhabited rural regions. They claimed that regional planning for land and water resources would alleviate farm poverty and restore the viability of rural living. One such "policy tributary" was a collection of Eastern planners who fought to extend electric service to farm areas. Rural electrification advocates such as Gifford Pinchot and Morris Cooke of Pennsylvania argued that the modern business and financial systems unnecessarily pulled impoverished people to overcrowded city centers. They maintained that the farmer's standard of living could be improved by bringing inexpensive power, especially hydropower, to the countryside. A group of government officials in the U.S. Department of Agriculture (USDA) formed the second policy tributary. Land economists such as L.C. Gray and M.L. Wilson believed that the agricultural depression of the 1920s reflected structural and environmental imbalances within farming. Other farm advocates argued that merely equalizing the farmer's terms of trade (making the tariff "effective" for agriculture) would correct disparities in income and living standards. But these planners insisted that the underlying problems stemmed not from the absence of a protective tariff or from the inability of farmers to organize collectively, but from the failure of agricultural systems to adapt to proper land use.

Though these groups only occasionally acted in concert, by the late 1920s and early 1930s their ideas coalesced into an influential body of work. Both the rural electrification advocates and the land-use planners linked social with environmental goals. They believed that rural poverty could be ameliorated by proper resource use and fair resource distribution, and they called upon the government to execute this vision. In this effort they enlisted as occasional allies Senator George Norris of Nebraska, a persistent advocate of public power and rural assistance, and Hugh H. Bennett, an equally relentless foe of soil erosion who campaigned against land degradation from his position in the Department of Agriculture. Because the same prescriptions for rural underdevelopment—cheap electricity, industrial decentralization, land planning, and land retirement—emanated from so many different quarters, these proposals received wide circulation and entered the political marketplace.

Presidential politics proved to be the critical arena. By the late 1920s, Herbert Hoover had earned his stripes as a sincere, if not fanatical, conservationist and farm advocate. He negotiated the first interstate water allocation compact in 1922, spearheaded efforts to conserve oil and timber, championed outdoor recreation initiatives, and supported federal outlays for flood control. He also believed that the federal government should assist the farmer by improving waterways, sponsoring and disseminating scientific research, and coordinating the work of agricultural marketing cooperatives. Initially, most of the new conservation advocates approved of Hoover's activities. These men moved in similar professional and social circles, they voted Republican, and they shared Hoover's faith in enlightened management and associational methods.[4] But they gradually grew disenchanted with the president's rigidity on the electric power and farm issues. While he steadfastly promoted interstate utilities coordination and multipurpose river development, Hoover remained firm in his opposition to government transmission of power or regulation of utilities to achieve area-wide coverage. This put the president on a collision course with those who desired more active government involvement in securing the fair distribution of electric current. Hoover also appeared insensitive to the rural distress of the early Depression. He supported continued farm research and marketing assistance, but held the line against interest-group solutions (raising prices directly) or rural rehabilitation schemes. He did know of marginal and exhausted lands, and even believed that the federal government should begin to take them out of production, but he did not believe the state had the authority or the funds to sponsor comprehensive adjustments in environmental and economic relations.

By insisting that the government never provide electrical service to the end consumer, and by asserting that farmer self-organization alone would correct economic disparities, Hoover underestimated the strength of the new conservation and cast himself as the movement's primary adversary. The president's failure to adopt its tenets, particularly in the face of economic collapse, provided a political opportunity for any rival willing to promise immediate assistance for rural Americans. Into this opening marched Franklin Roosevelt, who believed that restoring the viability of rural living meant directly addressing the relationship between rural resources and social conditions. This strategy was honed during his years as governor of New York, when he sponsored a series of initiatives to classify soils, to purchase and reforest marginal lands, and to bring hydroelectric power to the state's agricultural areas. Dismayed by high levels of farm desertion and by overgrown metropolitan centers, Roosevelt hoped that improved agricultural practices and urban amenities would extend the American standard of living to rural areas. "The ultimate goal," he explained, "is that the farmer and his family shall be put on the same level of earning capacity as his fellow American who lives in the city."[5] These ideas shaped Roosevelt's presidential campaign strategy, in which he appealed to America's "forgotten man" and promised direct assistance to those at the very bottom of the economic ladder. Convinced by an analysis that linked the Depression to the absence of farm purchasing power, Roosevelt hitched his wagon to the domestic allotment plan, to government generation and distribution of electric power, and to a new philosophy of government-assisted rural development. Conservationists, farm advocates, and land-use planners now put their efforts behind a new presidential candidate and a new political party.

GIANT POWER AND RURAL ELECTRIFICATION

Gifford Pinchot, former chief forester and principal spokesman for the "old" conservation, fired one of the first shots of the "new" conservation.[6] After he assumed the governorship of Pennsylvania in 1923, Pinchot convinced the legislature to approve a "Giant Power" survey of the state's water and fuel resources. The national demand for electricity in homes, businesses, and factories had risen dramatically during the first two decades of the century. The governor, like many engineers, public officials, and business representatives, saw tremendous possibilities in the introduction of powerful production systems and long-distance transmission lines. In addition to delivering abundant supplies of electricity, such technologies could ease load difficulties by enabling the "interconnection" of local, regional, and interstate power resources. "We have been slow to recognize that distance is a rapidly disappearing factor in public utility development," Pinchot declared. But these technological improvements also facilitated vast business combinations. In 1924, the seven largest holding companies controlled just under half of all the generating capacity in the country; by 1929, the number of holding companies with control over the same proportion of electricity dropped to three. Pinchot fretted over the increasing concentration in the power industry, but wanted Pennsylvania's consumers to reap the rewards of large generating plants and electrical interconnection. "From the power field," the governor predicted, "we can expect the most substantial aid in raising the standard of living, in eliminating the physical drudgery of life, and in winning the age-long struggle against poverty." He conceived the Giant Power Survey as part of a larger attempt to strengthen state supervision over the power industry, and to provide the nation (and the federal government) with an innovative regulatory model.[7]

To direct the Giant Power Board Pinchot appointed Morris Cooke, an engineer who had led the Public Service Commission's effort to reduce electric rates in Philadelphia. Cooke believed that the utilities discriminated against small and domestic users, and that they failed to extend electric service to areas where demand existed. Because Pennsylvania's bituminous coal deposits, not its streams or rivers, comprised the state's primary source of energy, Cooke and his associates developed an ambitious proposal to achieve statewide electrical coverage by locating large generating stations at the mines and by building a distribution system of high-voltage, long-distance trans-mission lines. The planners claimed that such an integrated system of pro-duction and distribution would eliminate waste, reduce freight costs, and lower prices. And they asserted that not only would this integrated, intercon-nected system assure low power charges, but those rates would also become standardized and less discriminatory. "Already the leaders in the industry are beginning to face the fact that it will be difficult to maintain differences in rates *for similar service and use* in one interconnected system," Cooke wrote to Pinchot. "The arguments for such differences become weaker and weaker day by day."[8]

Cooke and Pinchot soon combined this mission for reduced rates with a crusade to electrify the countryside. Cooke castigated the industry's slow and inconsistent efforts to bring electricity to the farm, and he persuaded Pinchot that the Giant Power Survey should be expanded to include investigations of rural electrification. To this end Cooke diligently gathered information on Pennsylvania's agricultural population, on rural demand, and on the potential costs of rural service. He also helped establish an experimental electrified farm in cooperation with Pennsylvania State College, and wrote to equip-ment manufacturers such as International Harvester and Frick Machinery hoping to interest them in developing more appliances for rural consumers.[9]

But Cooke and Pinchot saw far more than satisfied farm customers; they envisioned electrification as the basis for a new rural society. Pinchot claimed that the first economic revolution, based on steam power, had bypassed the farms and concentrated factories and workers in the cities. As a result, country life declined and small communities decayed. "Neither the farmer's produc-tive power nor his comforts and conveniences increased in proportion with those of other workers and investors," the governor declared. But electricity could be delivered many hundred miles away from where it was produced, thereby reversing the "social tendencies" of the age of steam. Pinchot and Cooke believed the government must take an active part in providing electric service to rural areas, for this revolutionary source of power would decentralize industry, restore country life, and "put the farmer on an equality with the townsman." They well understood that this vision constituted a significant extension of conservationist thought. Many farmers, Cooke remembered, supported Theodore Roosevelt's (TR) efforts to protect streams, waterfalls, and mines from corporate monopolization because they were simply "good citizens." "Now," he wrote, "the term conservation takes on an entirely new meaning, for it spells cheap power for the farm."[10]

What is more, the Giant Power designers intended these proposals to jumpstart the regulatory powers of the federal government. Neither Cooke nor Pinchot favored government ownership of the utility industry—they wanted to sidestep that question altogether—but they desired a strong federal commitment to reduce rates. While the Giant Power Board understood that Pennsylvania's electrical future hinged on coal, it forecast that the nation's electrical future would depend on hydropower. "Water power will last forever," wrote Philip Wells, Pinchot's Attorney General. "Its use will greatly prolong the life of our mineral deposits . . . And the great bulk of it remains within Federal jurisdiction." Therefore the Pennsylvania reformers based their state program on a piece of national legislation, the Water Power Act of 1920. This Act, the outcome of prolonged congressional debate, established the authority of the federal government to set the conditions of hydroelectric power development on navigable streams and for sites within the public domain. Furthermore, the legislation gave "preference" to states and munici-palities with proposals to construct and operate hydroelectric projects. But the Federal Power Commission, the three-person board charged with approv-ing power licenses, rarely met and did little but prevent the duplication of leases. Wells, Pinchot, and Cooke expected that the vigorous implementation of the Water Power Act on the state level (and applied to coal deposits) could inaugurate a new decentralized society, with particular benefits for rural peo-ple. This new society, they hoped, would inspire national action. "The Great State," Cooke affirmed, "is going to grow up out of a revivified agriculture and a re-inspiration in small town life and the utilization of these in placing the government of our individual states on a plane of effective social purpose . . . It is fortunately a law of growth that we don't need more than one conspicuous example in order to move the mass."[11]

The Pennsylvania reformers unrelentingly emphasized their vision of low rates and rural electrification in order to contrast Giant Power with "Super Power," a competing energy plan put forward by Secretary of Commerce Herbert Hoover during the same period. Like Giant Power, the Super Power proposals aimed to meet the growing demand for reliable electric service in the Northeast and along the mid-Atlantic seaboard. Hoover envisioned a regional system of interconnected electric lines that enabled the transmission of power in the same way standardized equipment allowed the interchange of railroad cars. Interconnection, the commerce secretary maintained, would allow isolated generating plants to abandon their individual power reserves and rely upon a central reserve. Hoover's goals—centralized production, waste reduction, and cheap power—were not unlike those of Pinchot, but the Giant Power proponents denounced Super Power as an attempt to grant the utilities all the monopolistic advantages of combination without requiring them to distribute those benefits to society. "Widespread interconnections do not necessarily give access to the cheapest power or spell low rates to the consumer," argued Cooke. The Pennsylvania reformers asserted that the country must follow the Giant Power example to lower rates and thwart a dangerous electric monopoly. "Giant Power," wrote Pinchot to each member of Congress, "proposes to break down and put an end to the present unfair discrimination in rates . . . Giant-power means regulation by the people, Super-power means control of the people by the monopoly."[12]

Hoover vehemently denied these charges, claiming that industrial stability and interstate coordination would indeed reduce costs and consumer prices. "Interconnection," he informed the Philadelphia Chamber of Commerce (a group that opposed Pinchot), "does not imply capital consolidation or the building up of great trusts. It implies the sale and resale of power from one utility to another." As an example for the type of interstate agreement he favored, Hoover touted the Colorado Compact of 1922, which apportioned the water of the Colorado River among six western states and cleared the way for the river basin's multipurpose development. He hoped that the Northeastern states would follow the same voluntary strategy to ensure the "fluid" transmission of electric power. Furthermore, Hoover maintained the States possessed sufficient authority to regulate the distribution of power generated outside their borders (a claim reformers disputed). The secretary of commerce recognized the difficulties posed by rural electrification, but claimed those difficulties did not justify any new regulatory initiatives because industrial cooperation would soon correct the situation. "The agricultural problem," as he wrote to Morris Cooke, "is one of first getting our primary system onto right lines."[13]

The advocates of Giant Power drew inspiration not from the Colorado Compact, which yielded few benefits during the 1920s, but from the Hydro-Electric Commission of Ontario, Canada. Set up by the provincial legislature in 1907, the "Hydro" distributed electrical power at cost from plants at the Niagara and St. Lawrence Rivers to communities in Southern Ontario. It constructed and administered the production and transmission systems, while the municipalities retained ownership of the operation. The Hydro not only delivered power to cities and towns, but also to smaller villages and farming communities. Canadian and American reformers believed Ontario's magic combination of low rates and rural electrification had inaugurated an economic revolution in the province. "One of the impressive points about the Hydro is the fact that all the small towns have access to power on relatively equal terms," Pinchot wrote in a special "Giant Power" issue of *Survey Graphic* in 1924. "Here industrial development is widely diffused and even small towns are on the same footing as larger centers." Journalist Martha Bensley Bruere agreed. She alleged that cheap electricity had diversified the economies of Ontario towns, raised incomes, and relieved farm women from drudgery. The Hydro, she wrote, had "checked the human tide toward the great cities, and created a land with no visible signs of poverty." It was this heightened emphasis on rural living standards that defined the new conservation. To be sure, these reformers still sought to eliminate waste by designing more efficient production and transmission methods, but this goal now served a larger purpose. Equity, they insisted, should be the outcome of the streamlined system; benefits should be distributed in accordance with social needs. Power—electric power, manufacturing power, financial power—need no longer be confined to the city center.[14]

A fellow contributor to the Giant Power crusade of the early 1920s was Senator George Norris of Nebraska. An ardent conservationist who had sided with Pinchot in the Hetch Hetchy controversy, Norris believed that the country's natural resources should be developed in multipurpose fashion for the benefit of ordinary people, not for the profits of private firms or monopolistic trusts. Norris likewise saw great promise in a "giant power" network of long-distance electric transmission lines. This vast system, he envisioned, would halt the growth of large cities and distribute electricity to small towns and rural areas. Though he did not share Pinchot and Cooke's optimism about the possibilities of regulation (Norris supported the government ownership of all resource development projects), he agreed that cheap power would strengthen farm communities and bring manufacturing to the country. No longer would the nation's farmers be forced to take jobs in urban centers, thereby inflicting permanent damage to the country by weakening its fundamental agricultural base.[15]

Norris also drew inspiration from Ontario Hydro and publicized Canada's experiment as part of his strategy to defeat a series of Republican-led efforts to dispose of the Muscle Shoals properties in northern Alabama. During World War I, the federal government had constructed two nitrate plants and a hydroelectric dam along the Tennessee River to manufacture explosives during the conflict and fertilizer afterward. In the early 1920s, the Harding and Coolidge administrations favored measures to sell or to lease these properties to private interests such as Henry Ford or the American Cyanamid Company. But the bills fell into the hands of Norris, who chaired the Senate Committee on Agriculture and Forestry. Norris claimed that these businesses coveted the completed hydroelectric facilities, and that they deviously disguised their primary interest in power by promising to deliver cheap fertilizer to Southern farmers. He also believed that the federal government should retain Muscle Shoals as the starting point in a more ambitious scheme to develop the entire Tennessee River basin. A trip to Ontario in 1925 confirmed Norris's view that a large public power system could be operated effectively and efficiently. In Canada, the consumption of electricity increased as power prices decreased; the people of Southern Ontario did not have to pay the cost of watered stock and private-utility propaganda. So Norris fought a two-front war; on one side he battled legislation that would have placed Muscle Shoals in private hands, while on the other side he fought to bring about a particular social vision. In the latter crusade he was supported by the Giant Power reformers, who certainly believed that any measures to lease hydroelectric sites such as Muscle Shoals must comply with the Water Power Act of 1920, preferably with terms to guarantee low rates and rural electrification.[16]

On both fronts Norris often sparred with the secretary of commerce. While Hoover remained on friendly terms with Pinchot and Cooke, he had no patience for Norris and his philosophy of government ownership. Yet Hoover was no proponent of laissez-faire. He wanted to see Muscle Shoals put to work for southern agriculture, southern manufacturing, and national defense. And in general he favored the public development of rivers and waterways, arguing that lower shipping costs would help farmers and that the federal government could save millions of dollars in interest charges by financing large works such as hydroelectric dams. But Hoover's commitment ended there. The secretary insisted that the government could never go into the power business; it could not compete with its citizens by distributing and marketing electricity to the end consumer. Essentially, Hoover concluded, a government that competed with its citizens by going into the power business violated the American commitment to economic freedom.[17]

Little came of Norris's Muscle Shoals proposal, or Pinchot's Giant Power scheme, or even Hoover's Super Power initiative during the 1920s. Coolidge, followed by Hoover, vetoed each public ownership bill that Norris wrote and the Congress passed; the Pennsylvania legislature rejected the Giant Power program not once, but twice; and utilities executives persistently disappointed Hoover by their reluctance to cooperate in securing electrical interconnection for Northeast. Of course the lack of immediate political success disappointed many Giant Power advocates. "Where are the engineers and statesmen," pleaded economist Stuart Chase, "to [bring about] ordered cities, impounded waters, tended forests, the sweep of great transmission lines, clean rivers, workshops planned with the dignity of cathedrals, and the end of grime and despair?"[18] But despite the appearance of failure, these reformers crafted an influential new conservationist agenda. By claiming that the country's natural resources should be protected and developed for farms and farm areas, they persuasively set forward an innovative set of social and environmental considerations. Conservationists now argued that inexpensive electric power and industrial decentralization would alleviate rural poverty and raise rural living standards. During the same period a group of agricultural economists from the USDA reached similar conclusions, though they approached the problem from a different perspective.

THE FARM CRISIS AND THE LAND UTILIZATION MOVEMENT

Bitter complaints from rural citizens and their political advocates punctured the celebrated prosperity of the 1920s. While industrial workers and city dwellers experienced rising wages and a dazzling array of new consumer choices, most farmers received an unwelcome taste of the Depression to come. Emboldened by high prices and government encouragement, they had expanded agricultural production during World War I, often pushing into marginal or arid lands where success was far from certain. Insurance companies and rural banks fueled this expansion, providing loans on dubious real estate and boosting land values in a speculative frenzy. In just three years, from 1916 to 1919, the nation's farm income almost doubled. But it dropped precipitously after 1920, when overseas demand for American agricultural products plummeted. Besieged by this postwar contraction, farmers found themselves caught between the low prices they received for farm products and the high prices they paid for nonfarm items. From 1919 to 1921, for example, the amount of corn required to purchase a gallon of gasoline increased tenfold. Though most commodity prices gradually rose again after 1923, the farm situation did not improve. The wheat market suffered from overproduction throughout the decade. Commercial farmers increased their vulnerability to economic shocks by borrowing heavily to purchase land and machinery. In addition, tenants and sharecroppers remained stranded on exhausted, marginal lands.[19]

In response some farmers' groups attempted to organize agricultural producers into marketing cooperatives. Many farm supporters saw such private actions as the means by which the highly individualistic farm sector could secure the advantages of efficient business organization and assert its collective interests alongside industry and labor. Most cooperative marketing initiatives proved unsuccessful; such voluntary schemes had difficulty controlling the farmers who failed to reduce acreage or store their products. In 1921, George Peek of the Moline Plow Company proposed that the federal government could resolve this "outsider dilemma" by raising farm prices directly. National legislation, he and his supporters argued, could restore the terms of trade to that of 1909–1914, a period when farmers felt satisfied with the exchange value between farm and nonfarm prices. The plan found concrete expression in the McNary–Haugen bills, which emerged as the focal point of congressional farm-relief legislation during the 1920s. The McNary–Haugen legislation proposed to raise domestic prices by selling surplus stocks abroad. This two-price system, its proponents believed, did not entail special protection for agriculture; it would simply extend the benefits of the tariff (already enjoyed by American businesses) to farmers. A newly formed agricultural lobby in Washington composed of the American Farm Bureau Federation (AFBF), other farm organizations, and several key members of Congress strongly favored the McNary–Haugen bills, but the legislation faced decisive opposition from the Coolidge and Hoover administrations.[20]

Other reformers criticized this export dumping. From a new institutional base within the Bureau of Agricultural Economics (BAE), an influential group of farm experts crafted a radically different set of solutions to the farm crisis. These men insisted that any effort to reorganize American agriculture should begin with internal adjustments on individual farms and within the agricultural economy. Building upon the example of the USDA's Extension Service, the BAE economists believed that concerted research and education efforts could help farmers maximize income, conserve natural resources, and plan cooperatively. To this end the BAE published farm statistics, developed price-forecasting models, and sponsored research in farm management, land tenure, and land use. The BAE supported the McNary–Haugen bills, but only as an emergency measure, not as a substitute for agricultural planning. And more than most McNary–Haugenites or the AFBF, which represented commercial farmers, the farm economists worried about inequalities within agriculture, and claimed that the application of social-science expertise could alleviate chronic rural poverty.[21]

Research and recommendations on natural resources constituted an essential part of this alternative approach to the farm crisis. Land economists in particular argued that improper land use created instability in the agricultural economy, and that comprehensive land utilization (land classification followed by land-use adjustments) would bring stability to rural areas. In 1922, the land economists found an institutional home within the BAE's Division of Land Economics. Under the direction of L.C. Gray, this division sponsored research and provided the leadership for the emerging land utilization movement, a movement that soon became an indispensable component of the new rural conservation.[22]

Perhaps influenced by his history teacher Frederick Jackson Turner, Gray interpreted World War I as the final closing of the frontier. Imprudent agricultural expansion onto poorer and drier lands had created a postwar legacy of abused resources and a marked rise in farm tenancy. "It seemed clear,"

Gray remembered, "that the continued increase of population would necessitate the occupancy of poorer and poorer lands and the operation of diminishing returns." Such an unfortunate scenario led Gray and his BAE colleagues to conclude that the era of unrestricted land settlement should be brought to a swift close. Land scarcity combined with population pressure required a cautious new approach to natural resources and agricultural development. The problem, as Gray explained, was not that the country faced an absolute shortage of land, but that different categories of land use (farming, grazing, forests) would now compete against each other. The country could no longer trust individual initiative or private enterprise to provide for future needs. Careful consideration of whether land ought to be used for crops, pastures, or forests—even towns, roads, or utilities—must therefore precede any new settlement and guide farm management decisions. This approach, Gray maintained, had the potential to stabilize agricultural prices, preserve natural resources, and repair the fabric of rural society.[23]

From the start, the land economists focused on the cutover timberland of the Lakes States. During the final decades of the nineteenth century, lumber companies had removed the white pines and hardwoods from northern Wisconsin and Michigan. State governments, land colonization companies, and other boosters expected that the cutover counties could become profitable farms, and between 1900 and 1920 settlers flooded into the region. But the new farmers did not produce very much for the market—their farms were small, they improved few acres, and they owned comparatively little livestock. The farm experts sharply criticized the process by which poor lands had been sold to hopeful farmers without providing them with guidance and institutional support. State governments, they concluded, should restrict settlement to lands best suited for agriculture and should purchase the remaining portions of the cutover for reforestation and recreational use.[24]

The process of land classification in the Lakes States began in Michigan with the 1922 Land Economic Survey, sponsored by Michigan State University and the state departments of agriculture and conservation. The survey assembled a statewide "land inventory" of data on soils, vegetative cover, tenure conditions, tax burdens, and trade areas. Though P.S. Lovejoy, the project's director, avoided the term "land classification" because he thought it politically contentious, the survey operated under the assumption that cutover lands would eventually be earmarked for one of three uses: farming (cropping or pastures), forests, or recreation. Wisconsin emulated Michigan's example, but proceeded past the inventory stage with legislation aimed at actual land adjustments. In 1927, a Forest Crop Law granted subsidies to individuals and counties for reforestation, and two years later the legislature passed a zoning law authorizing county boards to "regulate, restrict, and determine" the areas to be used for farming, forestry, and recreation. Several planners touted the legislation—the first of its kind in any state—as a model for a national land program.[25]

The concept of economic marginality guided both the Michigan and the Wisconsin surveys. Land economists believed that there existed certain classes of land upon which no farmer, however bright or hardworking, could achieve a decent standard of living. "Marginal" and "submarginal" farms simply did not yield returns commensurate with the amount of labor and capital invested. The land might be too rocky, for example, or the soil too thin or unsuitable in some other way. But this economic definition often blended with a sociocultural construction of marginality that justified critical judgments of character. Perhaps "marginal" farmers had no one but themselves to blame if they lacked the intelligence to avoid or to vacate unprofitable areas. "For the most part, there is little that can be done for these men," admitted Richard Ely, the paterfamilias of land economics. "Like other business men, farmers who have accepted the risk of operating the submarginal land must take their losses."[26] Though most agricultural experts probably held similar views, many were much more optimistic about possibilities for rural rehabilitation. To be sure, they believed that the least intelligent and the least adaptable farmers would have to find new employment. But they hoped something could be done for many who found themselves located on poor land or in risky areas without guidance or information.

One such risky region was the northern plains, where aridity—not tree stumps or stony soil—determined whether land was suitable for farming. Like the northern Lakes States, Montana experienced a wave of new agricultural settlement in the first decades of the twentieth century. Enticed by larger homesteads, boosters, and high levels of rainfall, migrants streamed onto the short-grass frontier and plowed up the virgin sod to plant acres and acres of wheat. But fortunes fell rapidly after World War I, when prices dropped and drought set in. An epidemic of tax delinquency, foreclosures, bankruptcy, and farm abandonment swept the state. Many farmers lost their land to banks and insurance companies only to become tenants on property they had once owned; wags replaced the slogan "Montana or Bust" with "Montana *and* Bust." As John Wesley Powell had predicted, small parcels of 320 acres might work well in the humid East but did not provide a secure foothold in the arid West. "We had to design a new pattern of occupation for Montana," remembered M.L. Wilson, a Wisconsin-trained land economist and the state's director of agricultural extension. "We had to find out how we could work out an adjustment here, and farm so as to stay."[27]

Wilson had moved to Montana in 1908, hoping to strike it rich himself. He and his business partner used crop data from the USDA's Bureau of Statistics to calculate that a single cash crop, flax, would yield the highest profits. They invested heavily in land and machinery, and borrowed money to cover the first year's operating costs. But the venture failed; dry weather and hot winds crippled the crop, and the bank repossessed their new 32 horsepower tractor. From this experience Wilson concluded that even educated farmers and highly capitalized operations might fail in the region's volatile environment. To succeed, farmers must have guidance and information based on thorough research. In 1911, Wilson went to work for the state's land-grant college, where he supervised its dry land demonstration farms and experimental plots, and soon accepted a job as Montana's first county agricultural agent. From his research and extension experience, Wilson concluded that cash cropping alone was likely to fail, and that farmers should shift large portions of their operations to raising livestock. More pastures and less plowing, he believed, would sustain Montana's rural communities.[28]

The widespread distress of the 1920s only strengthened Wilson's resolve to spread the word. He worked with the BAE, the experiment stations, and the extension service on studies of plains farming and rural poverty. Researchers and county agents toured existing farms and operated others for demonstration. They convinced many banks, which owned most of the foreclosed or abandoned land, to sell only to those farmers who agreed to follow the directions of the area's supervisors. The experts encouraged the consolidation of small tracts of land, arguing that only large, mechanized farms would guarantee sufficient incomes. The researchers also introduced techniques to reduce soil loss, and encouraged farmers to diversify by raising stock and growing forage crops on small, irrigated sections. The focal point of Wilson's efforts was the Fairway Farms project, a corporation of seven farms that Wilson established with his former economics instructor and BAE boss. Wilson and his partner wanted to demonstrate that tenants could become farm owners when provided with sufficient credit and expert guidance. Only by "adjusting the relations of farmers to the land," Wilson explained, could Montana's agricultural economy be rescued and its farms operated by the people who owned them. The Fairway experiment suffered large financial losses, but it demonstrated that enormous farms, with maximum machine power and very little human labor, could compete in the depressed world market. Wilson remained ambivalent about Fairway's implications, however, and continued to hope that the "old-time family farm" might not be swept away entirely.[29]

The poor social, economic, and environmental conditions of the cutover regions and the northern plains during the 1920s appeared to confirm L.C. Gray's contention that the best lands in the country had already been taken, and that settlements in marginal areas faced serious obstacles. For "the overexpansion of agriculture and the misdirection of agricultural expansion" Gray blamed boosters, real estate interests, transportation companies, and the federal government's "excessively liberal" land policies. "We have become so habituated to this characteristic of national development that we have achieved a comfortable cynicism which accepts it as the inevitable cost of progress," Gray observed in 1925. "One often hears that it 'requires three waves of settlers to develop a new country,' and so we are heedless of the waste in property and in lives."[30]

Developments in the field of soil science also strengthened the land utilization movement. Land classification and land planning decisions ultimately depended on whether a particular soil would sustain a particular agricultural activity, and from Hugh H. Bennett, a researcher at the USDA's Bureau of Soils, the land economists began to hear terrible news about the failure of agricultural practices to adapt to soil type and soil conditions. During the first part of his career surveying and mapping the soils of the South, Bennett became increasingly alarmed by the high rates of erosion he found. While gully erosion practically announced itself to observers because the land was so scarred, sheet erosion—the washing-away of thin layers of topsoil—threatened more farms by its very invisibility. Based on his observations of sheet erosion, Bennett concluded that decreased agricultural production and low incomes in some areas might be explained by this gradual, imperceptible loss.[31]

Armed with this new analysis Bennett launched a national campaign to publicize his work. With W.R. Chapline he coauthored *Soil Erosion: A National Menace* for the USDA in 1928, and he also wrote articles for popular magazines to reach a wider audience. Central to Bennett's message was the idea that proper land-use practices could halt soil erosion. He demanded that "every loyal citizen of the Nation lessen this tremendous evil" by building terraces and dams, planting grasses and trees on the unstable soils and sloping areas, and reducing the number of livestock on overgrazed regions. But Bennett also believed that the federal government should assist the individual farmer. He wanted to establish 18 new research and demonstration centers in regions particularly vulnerable to soil erosion, and he convinced several experiment station directors to assist him in this endeavor. Armed with examples of land degradation and alarming figures on lost farm income, Bennett testified before a Senate subcommittee and asked the Congress for money to begin work. Representative James Buchanan of Texas responded favorably, and he introduced legislation to begin the soil program. The final appropriation, approved by Congress in 1929, authorized erosion surveys, investigations of soil and water loss, and new conservation methods for erosion control. By 1931, eight experiment stations were in operation and ten more were in the works.[32]

During the 1920s, land economists and soil scientists concluded that improper land use resulted in unprofitable agricultural operations, tax delinquency, and farm abandonment. Their analysis, which linked the "farm problem" to the availability and use of natural resources, constituted an important contribution to the new rural conservation. Like the Giant Power advocates, these farm experts extended conservationist principles to inhabited rural areas. Land, they argued, should be protected and used correctly not just for its own or for efficiency's sake, but to raise the living standards of the people living on it. Furthermore, these planners found a partially receptive ear in the White House. Herbert Hoover and Secretary of Agriculture Arthur M. Hyde understood the concept of economic marginality and believed that the retirement of marginal land could eventually play a role in their federal farm program. But they did not embrace the social implications of land-use planning, or the political vehicle through which the land economists hoped to introduce it to rural Americans. In fact, Hoover's conservation policies in general, coupled with his increasingly rigid political philosophy, gradually alienated the new conservation advocates from his administration.

HERBERT HOOVER AND RURAL CONSERVATION POLICY, 1928–1932

Herbert Hoover entered the White House in 1929 with plans for a "new day" in American public life. He envisioned a country in which men and women enjoyed "the advantages of wealth, not concentrated in the hands of the few but spread through the lives of all," and where freedom from "poverty and fear" allowed opportunity for greater service to the community, the country, and the world. Hoover claimed the increasing complexity of modern life required the federal government's role in gathering information, in making "the fullest use of the best brains," and in assembling informed suggestions for the nation's improvement. But Hoover also insisted that these tools be used primarily to assist the voluntary activities of the people and their local or state governments. Too large or too intrusive a national state would threaten liberty by regimenting and dominating the country's economic life. Therefore the federal government might relieve poverty, guide prosperity, and preserve freedom by coordinating (and occasionally financing) the organization of those interested in collective self-help. This vision of state-sponsored voluntarism guided the president's approach to waterways development and the farm crisis.[33]

Hoover drew from his experience with the Mississippi flood of 1927 to illustrate his political philosophy and to set forward a national program of river development. Never had the lower Mississippi experienced such a deluge: during 2 weeks in April the river broke through levees in Missouri and Mississippi, killed over 250 people, and covered parts of 7 states—almost 16 million acres—with an inland lake of muddy water. President Coolidge appointed Hoover to direct the rescue, relief, and reconstruction efforts, and the commerce secretary did this to great national acclaim by coordinating the resources of the federal government with those of private agencies such as the Red Cross and the U.S. Chamber of Commerce. Hoover also helped publicize the U.S. Army Corps of Engineers' plans to improve flood control measures along the Mississippi, and he endorsed congressional actions to authorize large new federal outlays for this purpose. But Hoover saw an opportunity in the Mississippi tragedy to move beyond a single river and to advance a more aggressive agenda for the nation's waterways. "I am convinced," he wrote in 1928, "that we have come into an entirely new era in the development of our water resources . . . the flood serves to bring home to the American people the increasing dangers which lurk in our great streams if not adequately controlled." Hoover believed that the improvement and administration of waterways had once been simple: navigation had been the single concern and the task had fallen to one agency: the Corps of Engineers. But the modern economy and a growing population meant that flood control, irrigation, and electrical power development would become as important as navigation, and Hoover asserted that the federal government must take the lead in coordinating the work of public and private interests in those pursuits.[34]

Hoover argued for comprehensive, multipurpose development of the nation's water resources. Each drainage system—Mississippi, the St. Lawrence and Great Lakes, the Colorado, the Columbia, the Tennessee—should be studied with a view to understanding its potential for transportation, flood control, irrigation, and electric power. "What we need," Hoover explained, "is immediate determination of the broad objective and best development of every river, stream, and lake in order that we do not undertake or permit haphazard development, whether public or private, that will destroy the possibilities of their maximum development." But Hoover understood that the country faced "a problem of determining the limits of private and public development." He explained that the federal government had long financed navigational improvements due to its responsibility for interstate commerce, and it also contributed to flood control works on a cost-sharing basis with local and state governments. In addition, the federal government loaned capital for irrigation and reclamation projects with the understanding that the cost would be recovered. But Hoover admitted the problems became "more complex" when electric power was involved, and he stood by his opinion that the government should never go into the business of generating and distributing electricity. If a site were developed for power only, then it should be leased to a private party. When power was a "by-product" of government works constructed for navigation, flood control, or reclamation, the federal government should lease the power rights to recover the cost of the investment.[35]

This position explained Hoover's divergent approaches to the two most significant water power proposals of the 1920s: Boulder Dam and Muscle Shoals. The Boulder Project represented the culmination of Hoover's earlier work as chair of the Colorado River Commission, which had negotiated the interstate Colorado River Compact in 1922. Despite continued conflict over the compact's water allocation provisions, especially between California and Arizona, Congress passed the Boulder Canyon Bill in 1928. This legislation authorized the federal government to build a high dam on the lower river to protect the Imperial Valley from floods, to provide irrigation water and an "All-American canal" for the Southwest, and to furnish municipalities with hydroelectric power and a domestic water supply. But before the government could begin construction, it needed to secure contracts for the sale of power in order to pay for the project. Public power advocates wanted the federal government to operate the power facilities so as to expose the exorbitant rates of the power trust, but Hoover and his Interior Secretary Ray L. Wilbur held the line against such notions. They made it clear that the federal government would purchase the dam, tunnels, power house, and penstocks, but that the cost of the generating machinery would be repaid by the lessees. In addition, the lessees would build and operate their own transmission lines. "This will place the technical problems of generation and transmission of power in the hands of the purchasers," Wilbur explained. It also protected the boundaries that Hoover had drawn between public investment in multipurpose water projects and private enterprise: the federal government would not enter the electric power business. But in keeping with their belief that the federal government could strengthen local initiatives, Hoover and Wilbur did nothing to restrict smaller nonprofit groups from distributing current; in fact, the administration agreed to lease the lion's share of the power—over 90 percent—to public entities such as the Metropolitan Water District of Southern California, the city of Los Angeles, and other municipalities.[36]

Still, many public power advocates felt betrayed by this government "giveaway." The Boulder Dam power contracts, though they favored public entities, did not provide a way for regulators to force down private utility rates, nor did they establish any provisions to ensure that rural consumers received electricity. "Power progressives" such as George Norris were equally outraged that the administration agreed to lease 9 percent of the power to the Southern California Edison Company. Even this small amount of power, they felt, violated the preference clause embodied in reclamation law and the Federal Water Power Act. Norris, of course, favored the public ownership and operation of water projects, and he continued to sponsor Muscle Shoals legislation that would place the property in the hands of the national government. Coolidge pocket vetoed one such bill in 1928, but Norris hoped Hoover would approve a similar measure passed by Congress in 1930.

Like Norris, Hoover wanted to see the Tennessee river basin developed for flood control, navigation, power, and the production of fertilizer, but Norris's bill allowed the government to build transmission lines and to sell power to nonprofit public groups, especially farm organizations. The president believed the properties should be put into the hands of private interests, and censured Norris's bill in his veto message of 1931. "The power problem is not to be solved by the Federal Government going into the power business," Hoover declared. "I hesitate to contemplate the future of our institutions, of our Government, and of our country if the preoccupation of its officials is to be no longer the promotion of justice and equal opportunity."[37]

At issue was the definition of equal opportunity. For Norris, the idea meant that each citizen should have equal access to property he owned—the nation's natural resources. For Hoover, equal opportunity meant that the government should not hinder its citizens by competing with them. Instead, the president continued to recommend that the Muscle Shoals properties be leased or sold to "competent and experienced industrialists" who would produce and distribute fertilizer and power on a "cost plus, public service" basis. This would allow the government to recover the cost of federal investments in flood control and navigation, just as the power leases and irrigators' reclamation payments would pay for the Boulder Project. In stark contrast to the Boulder Project legislation, Norris's Muscle Shoals proposals included no provisions for repayment, and committed the federal government to produce and distribute electricity without the aid of private capital. Essentially, Hoover feared that by going into the power business, the government would enfeeble the nation's economic system by eliminating the incentives for individual effort and local self-government. And he wholeheartedly disapproved any scheme that might drain the federal treasury.[38]

Hoover's most unlucky scrap with the advocates of cheap hydropower occurred over the St. Lawrence River and with Franklin Roosevelt, the governor of New York. Throughout the 1920s Hoover had recommended the expansion of the "Great Lakes System" of inland waterways, arguing that a ship channel connecting the Lakes with the Atlantic would reduce transportation costs for Midwestern farmers and manufacturers. Such plans, however, depended on successful negotiations with Canada over the joint development of the St. Lawrence River. Into this diplomatic conflict entered Roosevelt, who insisted that New York's Power Authority take part in the discussions. Continuing a policy initiated by his predecessor Alfred Smith, Roosevelt had successfully reserved the state's water power sites—including that portion of the St. Lawrence within the state's boundaries—from unregulated private development. Like Norris, Roosevelt was impressed by the low rates and service record of the Ontario Hydro-Electric Commission. In 1931, he established the New York Power Authority to conduct studies on the transmission and distribution of low-cost current on the New York side of Niagara Falls, and he persuaded Morris Cooke of Pennsylvania's Giant Power Survey to join the Authority. Roosevelt staked much of his political and economic strategy on the development of New York's power sources, and called especially for rural electrification as part of a statewide plan to revivify farm areas.[39]

Roosevelt argued that New York's participation in the international negotiations was necessary to begin the project and to determine his state's share of the construction costs. "I am deeply interested in the immediate construction of the deep waterway as well as in the development of abundant and cheap power," Roosevelt wrote Hoover. "May I respectfully point out that such action would hasten greatly the initiation of . . . cheap electricity from the State-owned and controlled resource, to be developed for the primary interest of homes, farms, and industries." But Hoover insisted that the domestic distribution of power—the "by-product" of navigation works—was a secondary matter, and he rebuffed New York's request to participate.[40] What may have been the president's legally justified attempt to restrict international negotiations to the federal government appeared as a veto of cheap hydropower. By continuing to classify electric current as the "by-product" of other developments, Hoover sidestepped the question that gripped Roosevelt, Norris, Cooke, and many other rural advocates—whether any government, state or federal, had the authority to require that inexpensive electricity be distributed on an area-wide basis and at equal cost to scattered farm customers. Hoover insisted that solving the rural problem did not require government production and transmission of electricity or a vastly expanded regulatory system. Cheaper electricity, he convinced himself, would materialize as a result of prudent investments in waterways infrastructure and after private interests (not upstart state governors) made arrangements to distribute current with the public's interest in mind.

A similar worldview guided Hoover's approach to the farm crisis. Like the economists from the USDA's BAE, Hoover believed that farming suffered from internal weaknesses and that its rescue depended not on price-raising or export-dumping schemes, but on a program of government-assisted self-organization. Though Hoover's "baby"—the 1929 Agricultural Marketing Act—drew little enthusiasm from farm organizations or farm-state congressmen, it represented the first time any administration committed itself to a comprehensive farm policy. The legislation created a Federal Farm Board, an agricultural "action" agency whose members quickly decided upon a two-part plan: to modernize the farm marketing system by sponsoring centralized commodity associations, and to develop a cooperative and voluntary system of production control. These goals reflected the president's analysis of agriculture's ills; farming, he thought, lacked the kind of managers who had successfully organized the efficient systems of production and distribution in American industry. Hoover hoped that the Farm Board's efforts to build cooperative marketing and production control systems would create opportunities to fill this managerial void.[41]

Both the attempt to create centralized marketing structures and the attempt to institute a program of voluntary acreage reduction were unsuccessful. The board assumed that national commodity cooperatives could stabilize prices by holding products off the market, but as prices declined sharply after 1929 it became clear that the cooperatives could not remain solvent without large infusions of government credit. There was little hope that the cooperatives could become independent; it seemed that only continuous state assistance would allow them to act as market stabilizers. Additionally, the board itself began buying up certain quantities of wheat and cotton as its members increasingly framed the farm problem as one of overproduction. As a corollary to these stabilization operations the board tried to cajole farmers into reducing their acreage, but their efforts fell flat. In the end, the board was simply unable to resolve the tension between individual and collective interests—farmers could hope to reap the advantages of higher prices without themselves joining a marketing association or reducing their output.[42]

The faltering of Hoover's program and the deepening of the Depression opened the way for alternative proposals from the agricultural economists and land-use specialists of the BAE. Though they shared Hoover's commitment to building a more efficient farm sector through voluntary and non-bureaucratic methods, the economists did not share his faith in cooperative marketing. Farmers, they believed, did not suffer from an organizational or managerial void; what they lacked was information and competent technical guidance. Furthermore, neither Hoover nor the McNary–Haugen types considered the likelihood that the country had overinvested in agriculture, or that national policies should address the presence of a substantial underclass of poor and marginal farmers.[43]

L.C. Gray and BAE Bureau Chief Nils Olsen seized the opportunity to advance land-use planning as the proper response to the farm crisis. They linked poor land utilization to a problem that many of their colleagues, including the Farm Board and Hugh Bennett, now perceived as the primary difficulty facing U.S. agriculture: overproduction. Gray claimed that crop surpluses, low prices, and low farm incomes indicated the general maladjustment of American agriculture. "Either farmers are continuing production on lands no longer adapted to profitable farming," Gray claimed, "or they are attempting to maintain systems of farming which must be radically modified." Whereas in the early 1920s Gray had advised a cautious approach to farm development, by the end of the decade he urged the country to "liquidate" the results of agricultural overexpansion. National land-classification efforts and a "discriminating extension policy," Gray believed, should encourage the abandonment of poor land and promote agricultural reorganization in areas fit for continued occupancy. In addition, Gray maintained the government should launch programs to purchase submarginal land and guide displaced farmers to new agricultural or industrial employment. Gray and Olsen sought greater publicity for their ideas, and befriended Arthur Hyde, Hoover's secretary of agriculture. Although Hyde was reluctant to endorse a full-fledged BAE land-classification program, he did agree to sponsor a national land utilization conference in the fall of 1931.[44]

The conference, held in Chicago and organized jointly by the Department of Agriculture and the Association of Land Grant Colleges, crystallized a decade's research in land utilization. Academics, government employees, and delegates from several farm organizations presented papers to 350 attendees, including their university and government colleagues as well as representatives from chambers of commerce, railroads, banks, and insurance companies. The policy recommendations largely reflected Gray's influence: a land inventory, the mapping of crop and range areas by soil regions and economic returns, the public acquisition of marginal land, the cessation of new reclamation projects, and improvements in agricultural credit and tax codes. The attendees also recommended the establishment of two national committees: a Land Use Planning Committee to coordinate research and formulate long-term planning policy, and an Advisory and Legislative Committee on Land Use to submit legislation to Congress.[45]

Olsen and Gray hoped that these Land Use Committees would inaugurate a new era of permanent agricultural planning, and Gray's Division of Land Economics threw itself into the preparation of many of the research reports issued by the committees in 1932. "These releases," Gray remembered, "preached the doctrines that the long reign of individualism had created widespread social and economic maladjustments in the use of land, [and that] each acre of land has a socially best use, which must be discovered through the process of land planning." But the endless research did not bring about a comprehensive new land policy or a new system of agricultural planning. Rather than crafting legislation giving farmers incentives to cooperate, the committees continued to recommend expanded programs of agricultural extension and technical guidance.[46]

However, not all farm and land utilization advocates were at a loss for ideas. In particular, former BAE employee and Montana State economist M.L. Wilson entered agricultural policy circles with a set of proposals designed to break the impasse. As the Depression deepened, as farm prices dropped, and as international trade fell victim to fits of economic nationalism, Wilson concluded that the problems facing American farmers demanded a two-part solution: first, emergency measures designed to reduce domestic crop surpluses and to raise farm income immediately; and second, the implementation of agricultural planning machinery to bring about proper regional adjustments over the longer term, including the abandonment of marginal land. Wilson now turned his attention to the promotion of a "Voluntary Domestic Allotment Plan," a measure he believed had the potential to bring about both short-term and long-range goals.[47]

The domestic allotment plan, which proposed to raise commodity prices without dumping crops abroad and without raising domestic production levels, entered farm policy discussions in the late 1920s as an alternative to the McNary–Haugen bills. The plan would require processors of agricultural products, such as flour millers, to purchase certificates issued to farmers on the basis of their "domestic allotment," or the share an individual farmer contributed to the total amount of a commodity bought and consumed within the country. Wilson became intrigued by the allotment proposal because it offered immediate assistance and could therefore serve as the first phase in a program of agricultural recovery. But because the allotment plan might bind farmers even more closely to their existing crops, it would be necessary to join it with plans for long-term agricultural and land-use adjustments. In his subsequent quest to publicize the allotment plan, Wilson found a close ally in Mordecai Ezekiel, an economist with the Farm Board. Like Wilson, Ezekiel had concluded that the abysmal state of international trade demanded a purely domestic approach to the farm problem: emergency measures to reduce crop surpluses coupled with long-term agricultural planning. Under Wilson and Ezekiel, the allotment plan metamorphosed into a "smokescreen" for the "heavy artillery on agricultural planning." A processing tax would replace the certificates, thereby providing the federal treasury with money to pay farmers willing to reduce production voluntarily. Representatives from the land-grant colleges and the Extension Service would participate in the administration of the program, but only through local committees composed of farmers, bankers, and community leaders. Wilson theorized that this decentralized committee system, with its potential for local and democratic decision-making, would provide the foundation for long-term shifts in local and regional agricultural strategy.[48]

The central component of Wilson's long-term strategy was a national program of land utilization. Wilson agreed with Gray that the nation's unrestricted homestead policies had created human and environmental maladjustment in rural America. The federal government, Wilson argued, should sponsor a national land inventory and should assume the leadership in a program of land-use research and policy. Programs in mountainous areas and in cutover regions might include reforestation, wildlife and game management, and recreational activities, while programs in the desert and public domain states might address rangeland rehabilitation, vegetative cover, and stock water development. Like Gray, Wilson also recommended that the federal and state governments purchase submarginal lands. Such acquisitions, he claimed, would consolidate public holdings for forests, parks, and watershed protection, and would remove from private ownership lands that could not be used profitably without human hardship and soil degradation. As an example of land utilization in action Wilson pointed to the program in New York State, where Governor Roosevelt had launched an ambitious land-classification and land-purchase scheme. But Wilson went further than Gray in an attempt to link these new land proposals with concrete measures to discourage crop surpluses and to balance production with domestic demand. Wilson understood that the problem of overproduction was not simply confined to marginal land, but that surpluses were just as likely to emanate from good land and from farms and farmers considered "supermarginal" in the terminology of land economics. Therefore Wilson insisted that some form of the domestic allotment plan should accompany the land program.[49]

Like most farm economists, Wilson looked forward to larger farms, fewer farmers, and the revival of overseas markets. But such a vision depended on employment opportunities for displaced farmers, opportunities sorely lacking during the Depression. Would the farmers released from poor agricultural lands, he asked, join the already overcrowded ranks of the urban unemployed? Wilson's answer was a firm negative. Agricultural and industrial policy could both rehabilitate American farming and restore "balance" to the rural–urban relationship. "There is a twilight zone between industry and agriculture," he wrote, "where higher standards of living are available which are consistent with industrial employment on the one hand and better use of the land on the other." Echoing the Giant Power proponents, Wilson argued that industrial decentralization could bring manufacturing jobs to the country, and that small-farm "subsistence homesteads" could combine factory work with part-time agriculture. Another possibility was the resettlement of displaced farmers onto viable lands, possibly even onto well-planned and self-financing irrigation projects.[50]

Wilson and Ezekiel purposely combined such land-use adjustment proposals with the domestic allotment plan to win over the more traditional farm organizations, and they often employed the language of parity to defend what they hoped would be a temporary measure to raise farm income directly. "Land-use planning must be the foundation for any farm-relief program," Wilson declared in a 1932 radio address. "At the same time, neither this nor any combination of plans can be successful with a wobbling monetary system that forces farmers to pay their obligations with three times the amount of farm produce that the obligations represented when they were contracted." Coupling the two programs, however, lost them the president's support. Wilson and Ezekiel attempted to secure Hoover's approval, but he expressed great disdain for the domestic allotment plan. Agricultural recovery, Hoover claimed, depended not on interest-group legislation, but on the generous expansion of short-term credit such as that provided through the Federal Reserve, the new Reconstruction Finance Corporation, and a reorganized system of federal land banks. "There is no relief to the farmer by extending Government bureaucracy to control his production and thus curtail his liberties, nor by subsidies that only bring more bureaucracy and ultimate collapse," Hoover declared before the Republican Convention of 1932.[51]

Yet Hoover was somewhat amenable to the goals of the land utilization movement, and he approved a land-leasing measure put forward by Secretary Hyde in response to the domestic allotment plan. Hyde proposed that the federal government levy a small processing tax to lease marginal lands, with an option to purchase. "This would fit into a general land utilization program," Hyde argued, "whereby the government eventually would take over much of the poor land now being used for raising crops and plant it to trees or convert it into public parks." Hyde claimed that the leasing proposal was "less harmful" than the domestic allotment plan because it would cost less, and would not require intricate bureaucratic machinery. He also maintained that the least productive and most marginal land would be automatically retired, as farmers making the least profits would submit the lowest bids. "The low cost producer on fertile lands would not be disturbed," Hyde explained, "[and] this is as it should be." But by 1933 most farm advocates agreed that the low-cost producer on fertile lands was part of the overproduction problem. Hoover and Hyde resolutely refused to entertain Wilson's contention that farm policy required two phases: first, emergency measures to restore balance between the nation's supply and demand for farm products (paying all farmers to reduce crop surpluses); and thereafter, steps to build a long-range program of land planning and agricultural adjustment. Wilson, for his part, dismissed the leasing proposal as a conservative measure backed by millers and food packers who wanted to avoid the processing tax; he had little faith in Hyde's assurance that the Hoover administration would eventually purchase marginal land and put it to proper "public use."[52]

Nor did Hoover and Hyde suggest what might become of the displaced farmers. The government, they said, would naturally allow the families to remain on the leased land and to raise food and "garden stuff" for their own use. But was mere subsistence a permanent solution? If, as Hoover conceded, the federal government should begin to take marginal land out of production, he never acknowledged Wilson's corollary—that the government must also create alternate employment opportunities in rural areas. Hoover, in fact, offered scant support for the single "resettlement" proposal of the early

Depression: a scheme for planned farm communities in the South. A collection of southern congressmen, businessmen's associations, and academic sociologists (as well as the Reclamation Bureau, the sponsoring agency) argued that a few demonstration communities might indicate how southern cotton farmers, especially tenants trapped by a single crop on exhausted land, could improve their conditions. As if on cue, Hoover indicated that such programs were best left to local initiative, perhaps with a small amount of financial assistance from the Farm Board.[53]

Despite his engineering and humanitarian credentials, Hoover disappointed rural advocates and became the primary impediment to the new conservation. His approach to the electric power and the farm relief issues revealed that Hoover thought industrial organization and private (but public-spirited) cooperation would automatically benefit rural Americans. Cheaper electricity, efficient farming, and proper land use would somehow flow out of the streamlined system. To sell power to the final consumer, or to raise the price of farm products directly would bankrupt the government, morally and financially. Hoover's prescriptions, however, were at odds with the latest developments in conservationist thought. Reformers such as Morris Cooke, George Norris, and Franklin Roosevelt did not agree with Hoover that abundant, inexpensive electricity would inevitably flow to the farm from a more efficiently organized utility industry. And other planners such as M.L. Wilson and Mordecai Ezekiel did not believe that individual farmers, pinned between low commodity prices and few employment alternatives, would be able to cooperate with a national land utilization program without tangible incentives (and the administrative capacity) to do so. One must conclude that in the end Hoover simply disagreed with the bedrock principle of the New Conservation—that the farm problem was directly traceable to maladjustments in resource distribution and resource use. If he had agreed, Hoover might have regarded rural conservation measures as one way out of the Depression, rather than as a reward to be earned in the event of citizen self-organization and economic recovery.

FDR: GOVERNOR AND CANDIDATE

Franklin Roosevelt entered the presidential campaign of 1932 with a considerable background in the new rural conservation. While governor of New York, he responded to the state's astonishing rate of farm abandonment—approximately 300,000 farms ceased operations during the 1920s—with a set of proposals designed to stabilize the agricultural population by means of better conservation planning. Roosevelt claimed that improper land use was the primary economic reason for the relative decline of the state's rural areas. "We have come to realize," he explained before a Cornell audience in 1930, "that many thousands of acres in this State have been cultivated at a loss, acres which are not under modern conditions suitable for agriculture." In addition, Roosevelt continued, New York had "used many thousands of acres for growing crops unsuited to the particular soil." Rural areas also faced unfairly high taxes, inadequate marketing processes, poor schools, poor health care, and the absence of a "socially interesting" farm life. Roosevelt argued that the problem therefore demanded a two-part solution: reforesting acres unfit for cultivation, and helping those engaged in farming suitable land to remain under more favorable and more profitable conditions. Regional planning, he explained, could "from the economic side make possible the earning of an adequate compensation, and on the social side the enjoyment of all the necessary advantages which exist today in the cities."[54]

The first farm legislation passed during Roosevelt's term as governor reduced rural taxes, and shifted the cost of schools and roads from the local to the state government. This tax relief served two purposes: it reduced the financial burden on farm communities, and it gave the state government more control over the process of regional planning. Roosevelt and his agricultural advisers also requested prompt completion of the state's land and soil survey. "When the study is completed," Roosevelt informed the legislature, "accurate data will be at hand to indicate definitely which lands of the State can profitably be continued in cultivation, which lands of the State should be devoted to reforestation, and which lands of the State should be used for industrial purposes." The land survey built upon work conducted in Tompkins and Chenango counties by Cornell professor and land economist G.F. Warren. By the mid-1920s, Warren had already come to the conclusion that farms on unproductive soils could not compete with farms located on better land or with city industries, and like Roosevelt he recommended that the state encourage farming on good land and forestry and recreation on poor land. The state legislature and New York voters subsequently approved a constitutional amendment authorizing a multimillion dollar bond issue to finance the reforestation of abandoned farmland. While most of the purchased land would be earmarked for future timber supply, the forests within the Adirondack and Catskill preserves would remain untouched. At the higher elevations, Roosevelt argued, the restored forests would also regulate stream flow, help prevent floods, and provide more dependable sources of water for villages and cities.[55]

New York's land program formed a critical part of Roosevelt's larger regional vision, especially after the onset of the Depression. He hoped that urbanites might find relief from hunger, joblessness, and congested conditions by moving into the countryside, and that the agricultural population once residing on marginal lands might locate employment within the state's better farming regions or in small industrial plants established in rural areas.

"Hitherto we have spoken of two types of living and only two—urban and rural," he explained. "I believe we can look forward to three rather than two types, for there is a definite place for an intermediate type between the urban and the rural, namely, a rural-industrial group." To be sure, this picture of small factories nestled into the upstate hills reinforced Roosevelt's romantic attachment to the back-to-the-land movement and justified his proposals to provide unemployed city dwellers with subsistence homesteads in the country. But his commitment to a more balanced distribution of population translated into concrete efforts to decentralize the state's economic resources and to raise rural incomes. "Remember please that land and its proper use is still the basis of the prosperity of a State," Roosevelt reminded his fellow New Yorkers in 1931. "I want to build up the land as, in part at least, an insurance against future depressions."[56]

Roosevelt also began his gubernatorial term with a promise to provide New York with inexpensive electricity by means of state-developed hydropower facilities on the St. Lawrence River and with a strengthened program of utility regulation. "In the brief time I have been speaking to you," he declared in his inaugural address, "there has run to waste on their paths to the sea enough power from our rivers to have turned the wheels of a thousand factories, to have lit a million farmers' homes." He pledged that the government would forever retain title to the state's power sites, and that electricity would be distributed at the lowest possible cost. Roosevelt left open the possibility that private enterprise might play a role in the distribution of the power, though he warned against a "too hasty assumption that mere regulation is, in itself, a sure guarantee of protection of the interest of the consumer."[57]

A few months later the governor followed up his cautionary advice with an admonition aimed directly at the utility companies: his administration would consider public transmission and distribution of the power if private interests proved unable or unwilling to provide service at reasonable rates. Roosevelt also suggested to a national audience in *Forum* magazine that public generation and distribution of hydropower in key locations such as Muscle Shoals, Boulder Dam, and the St. Lawrence River might serve as a useful "yardstick" with which "to measure the cost of producing and transmitting electricity" in general. Underlying the governor's power proposals was his belief that private companies discriminated against domestic consumers and small users. Roosevelt compared New York's electric rates unfavorably with those just across the border in Canada, where the provincial legislature had set up a public commission to produce and distribute hydroelectric power at cost to the farms, villages, and towns of Southern Ontario. Roosevelt, however, brandished the Ontario rates as an example of area-wide coverage at reasonable rates, not yet as an argument for public distribution as well as public generation. He still wanted to give the utility companies a chance to cooperate.[58]

In 1931 the state legislature passed an act establishing the New York Power Authority, and charged it with the task of developing hydroelectric energy from the St. Lawrence. The authority was to build dams and power-houses, and to make agreements with private firms for the transmission and distribution of the electricity. Power rates would be fixed by contract to ensure that rural and domestic consumers received inexpensive electricity at nondiscriminatory prices. Roosevelt appointed several experienced utility regulators to the Power Authority, including Morris Cooke of Pennsylvania. Like Roosevelt, Cooke claimed that private companies overcharged small users, and he argued that lower power rates combined with area-wide cover-age would increase the consumption of electricity. He concluded that just one successful example of public regulation at the state level might well inspire similar experiments across the nation, and he traveled to New York excited at the prospect of demonstrating once and for all how "regulation by contract" could bring about fair retail rates.[59]

The Power Authority at once began to conduct technical studies on how best to generate and distribute electricity. Its members and staff began this process with a willingness to cooperate with private industry, but they gradu-ally altered their views because one newly formed company, the Niagara-Hudson, controlled much of the state and occupied a monopolistic bidding advantage. Negotiations with Niagara-Hudson proved unsuccessful, and Cooke suggested that the state consider building its own transmission lines. He and other members of the authority also recommended to Roosevelt in late 1931 that municipalities and farm cooperatives be permitted to construct their own distribution systems. "It is hard enough to make regulation work," Cooke wrote in response to an inquiry about the status of New York's power program. "Therefore anything like this which tends to put the bargaining position on somewhat equal terms should have the support of the companies because it has become clear that public ownership is inevitable if regulation breaks down." Early the following year the governor asked state lawmakers to approve a law allowing municipalities to form utility districts, but the legisla-ture turned down this proposal. Roosevelt's plans suffered an equally devas-tating setback when the Hoover administration rebuffed his attempts to involve the Power Authority in diplomatic negotiations with Canada over the development of multipurpose works on the St. Lawrence.[60]

Despite the failure of Roosevelt's regional program to materialize during his term as governor, he never abandoned his vision of a rural renaissance built upon the proper use of land and the availability of inexpensive hydroelectric power. The Depression, in fact, only strengthened these commitments. Unlike Hoover, who insisted that farm recovery depended on general economic recovery, Roosevelt wanted to raise rural living standards directly. During the presidential campaign of 1932, he embraced rural conservation policy as a weapon in the struggle for economic recovery and as a political device to distinguish him from his competitor. Though Roosevelt recognized that no single factor would bring immediate prosperity to the agricultural population, he linked the Depression to the absence of farm purchasing power and claimed that low rural incomes kept factories idle and urban workers unemployed. "Our economic life today is a seamless web," he explained. "If we get back to the root of the difficulty, we will find that it is the present lack of equality for agriculture." Though he saw the need for "quick-acting remedies" such as surplus reduction, he claimed that permanent farm relief required national agricultural planning in general and land-use planning in particular. Roosevelt also viewed rural electrification as part of this overall strategy to restore buying power to the farmer. He claimed that the country had been unable to reap full advantage of its natural resources because selfish utility interests refused to establish rates low enough to encourage widespread electrical consumption. "We are backward in the use of electricity in our homes and on our farms," he declared. "Low prices to the domestic consumer will result in his using far more electrical appliances than he does today."[61]

A small group of academics and advisors known as the Brains Trust helped Roosevelt formulate this political and economic strategy. During the spring of 1932, as Roosevelt was preparing for the presidential race, the members met regularly at the governor's request to hammer out a set of organizing principles for the campaign. Raymond Moley, the group's founder, recalled that as of April no "Roosevelt program" existed, only the governor's "policies, near-policies, and mere leanings." The governor's power program, Moley reported, was foremost among these interests. Roosevelt not only claimed that the sources of water power belonged to the American people in perpetuity (a position taken by his uncle Theodore), but he also insisted that the government had a duty to see that the power was produced and distributed to the people at the lowest possible cost. In addition, the candidate's advisers understood that his power plan was only part of a larger vision of rural rehabilitation and regional development. "Roosevelt," Moley remembered, "had advocated reforestation, land utilization, the relief of farmers from an inequitable tax burden, and the curative possibilities of diversifying our industrial life by sending a proportion of it into the rural districts." Though Roosevelt was vaguely interested in short-term farm relief measures such as the McNary–Haugen plan, he impressed his advisers with the idea that long-term agricultural prosperity was linked to the use and availability of land and water resources. Roosevelt, Moley emphasized, essentially saw "the central problem of agriculture—the paradox of scarcity in the midst of plenty" as a problem of conservation.[62]

Accordingly, the Brains Trust listed "Agriculture" (under the larger heading of "Conservation") as the campaign's top priority. The group knew that

Roosevelt intended to address the nation's farm crisis, and that he also hoped to tempt the discontented Republican yeomen of the Midwest across party lines. But their candidate had not yet linked his desire to win votes, to relieve human misery, and to bring about a permanent rural life with a more general approach to fighting the Depression. Therefore his advisers recommended that Roosevelt repackage his inclination to address the country's agricultural troubles. The Depression, they argued, was not simply industrial in character; it had begun in the farm sector. It followed that national economic recovery depended on agricultural recovery, and that programs to assist the farmer directly would help the nation as a whole. "The obvious beginning of our discontent in this country," Moley explained, "was the persistence of the delusion that the nation could prosper while its farmers went begging."[63]

The primary intellect behind the campaign's farm strategy was Rexford Tugwell, a young professor at Columbia University. Tugwell, whom Moley described as a "first-rate economist who had pushed beyond the frontiers of stiff classicism," was an institutionalist who believed that planning (or "social management," a term he preferred) could shape economic and technological forces into instruments for the common good. Tugwell argued that the Depression originated in the domestic economic imbalances of the preceding decade. Production costs had fallen rapidly, and workers produced more per man-hour than ever before, but businesses failed to pass on these gains in efficiency as higher wages to workers or as lower prices to consumers. Instead, they retained the profits as protective reserves or built more plants. The result, Tugwell reasoned, was an imprudent increase in productive capacity (overproduction) without a commensurate increase in purchasing power (underconsumption). At a particular disadvantage were the nation's farmers, who suffered from the "deranged" exchange relationship between farm and industrial goods, and who lacked sufficient purchasing power to buy their share of manufactured products. City workers, in turn, found themselves on the unemployment lines when the factories shut down. Because urban unemployment was related to low agricultural incomes, Tugwell believed that the government should reestablish the "exchangeability" among groups. "It was a downright economic necessity that [farmers] should be consumers," Tugwell declared at the Roosevelts' supper table. "To keep the economy going," he insisted, "everyone must be able to buy and sell to everyone else."[64]

Tugwell, like Roosevelt, criticized the Hoover administration's recovery program for dispensing emergency aid only to financial institutions and production facilities at the top of the economic pyramid. He therefore looked approvingly upon Roosevelt's promises to provide assistance to the bottom as well as the top. And largely due to Tugwell's influence, Roosevelt began to justify farm relief as part of a general approach to economic recovery. "One-half of our population, over 50 million people, are dependent on agriculture," he declared after accepting the Democratic nomination. "And, my friends, if those 50 million people have no money, no cash, to buy what is produced in the city, the city suffers." Roosevelt urged that steps be taken to restore the purchasing power of rural Americans, and that the country accept a new picture of group cooperation. "This Nation is not merely a Nation of independence," he affirmed. "If we are to survive, [we are] bound to be a Nation of interdependence—town and city, and North and South, East and West."[65]

But Tugwell not only helped Roosevelt formulate this theory of rural underconsumption; he also shared the governor's (and the new conservationist's) conviction that the farm crisis was ultimately a problem for natural resource policy. Echoing the analysis of BAE economists such as L.C. Gray and M.L. Wilson, Tugwell argued that the country had not yet cultivated a permanent agriculture; most of the nation's farmers had not yet progressed beyond a pioneer stage of primitive and transient cultivation. World War I had further encouraged irresponsible expansion, leading directly to the agricultural depression of the 1920s. Individual interests, Tugwell claimed, had run contrary to the social interest, resulting in soil erosion, butchered forests, rising rates of tenancy, and widespread rural depression. To make things worse, none of the farm relief proposals in Congress required any actual change in the practice of farming. "Agriculture was to be made profitable," he complained in 1929, "and this was to be done uncritically and with no attempt to gauge the future or to penalize inefficiency or anti-social techniques." Rather, Tugwell insisted that the federal government should perform two functions as part of any farm program: it should withdraw marginal land from production and convert it into public forests, parks, and grazing reserves; and it should require that farmers maintain "continuous productivity" on private land by adopting improved practices such as plowing on the contour, planting more legumes, and replacing erosion-inducing row crops with grasses and trees. The fact that farmers were begging for relief, Tugwell suggested, gave the "expert his chance" to couple long-term conservation and land-use adjustment measures with financial assistance. "We had been mining our soil for a long time," he expressed to a sympathetic Roosevelt, "but we were now doing something even worse, we were exploiting our people, using up their savings and their capital goods and reducing their levels of life." Tugwell advised the presidential hopeful that the desperate situation was now his opportunity.[66]

As for a specific agricultural program, both the candidate and his Columbia adviser believed that it should combine production control with the nation's interest in conservation. Neither Tugwell nor Roosevelt approved of the price-raising and export-dumping proposals in circulation, for these would have subsidized producers without bringing about necessary changes within the practice of farming. Restricting production in the midst of widespread depression, Tugwell argued, was not as heartless or as half-baked as its detractors made it sound. It would not involve a net decrease in food supplies, but shifts in land use: replacing the soil-destroying staples (cotton, corn, and wheat) with soil-building grasses, soybeans and alfalfa. Still, Roosevelt found it necessary to court the major farm organizations, and he suggested to the American Farm Bureau Federation, the National Grange, and the Farmers' Union that he would accept any proposal they agreed upon together. However, behind the scenes he encouraged Tugwell to search for a better plan, one that combined production control and conservation with direct assistance so that compliant farmers might see their incomes rise immediately.[67]

Through Mordecai Ezekiel, Tugwell heard of one proposal—the domestic allotment plan—that met both of those conditions. He traveled to a farm economics conference in Chicago to meet with M.L. Wilson, the plan's chief publicist, and to hear from other proponents such as Henry A. Wallace, an influential agricultural editor, and Howard Tolley, an advocate of agricultural planning who had worked with the BAE. Tugwell was impressed with Wilson and with his specific set of policy proposals, and he convinced the Montana professor to book a ticket for Albany, New York, to meet with the Roosevelt camp.[68]

Wilson and Ezekiel had begun their efforts to publicize the domestic allotment plan in a spirit of neutral bipartisanship, but they faced opposition from the Hoover administration, which held the line against raising prices directly. They also drew the criticism of farm leaders, who wanted some combination of inflation and increased prices without meddlesome administrative interference. So by his July conference with the Democratic candidate, Wilson was ready with a new political pitch. The meeting went quite well; the governor especially warmed up to the professor when Wilson praised New York's land program. They also found themselves in total agreement about the need for submarginal land retirement and industrial decentralization. "Well now, M.L., let's get straight on this," Roosevelt said, slapping his knee, "have you got my plan, or have I got your plan?" But Wilson warned that a land program would go only so far. Farmers, he explained, were in such a state of mind that they wanted something done quickly. Midwesterners, traditionally Republican, were now ready to vote for a Democrat if he offered quick relief. Wilson insisted the governor could win those farm votes by promising aid without being too specific.[69]

Wilson's advice found concrete expression in Roosevelt's September farm address. The product of more than 25 people, including Wilson, Tugwell, and Wallace, the speech basically outlined the allotment plan without mentioning its name, without calling for acreage reduction, and without even outlining a specific method of farm payments. Nevertheless, Roosevelt deftly allied himself with the farmer by once and for all removing the taint of special-interest legislation from the idea of agricultural assistance. He alleged that poverty existed in the midst of abundance: farmers struggled just within reach of incomparable natural resources. And at the heart of the Depression was the farmer's falling share of the national income. Without his buying power, Roosevelt claimed, factories would remain closed and workers would remain idle. "This Nation," he declared, "cannot endure if it is half 'boom' and half 'broke.' " He called for permanent measures such as a program of national agricultural planning, a national policy of land utilization, and efforts to decentralize industry. However, Roosevelt granted that these "slow-moving" proposals had to be complemented by more "quick-acting remedies." An emergency farm plan, he said, must give benefits to agriculture that were equal to the tariff protection enjoyed by industry. Additionally, any such legislation should increase prices without increasing production; it should finance itself; and it should be voluntary. Wilson was thrilled with Roosevelt's sly approach. "Curiosity is a great means for developing and introducing ideas," he wrote Ezekiel. "It will be very easy for him after election to announce that the voluntary domestic allotment plan as embodied in such and such a bill completely fulfills his expectations."[70]

Roosevelt had less desire for such stealth when formulating a position on electric utilities and rural service. As Moley indicated, the governor had lavished most of his attention on this issue, developing a combative regulatory philosophy that coupled TR's emphasis on public ownership of waterpower sites with a new insistence on widespread distribution and nondiscriminatory pricing. Roosevelt had furthermore indicated that government-owned hydroelectric plants might usefully serve as a "yardstick" for setting private rates, and (unlike Hoover) he supported George Norris's attempt to include long-distance transmission authority within the Muscle Shoals legislation. Like the new conservationists, Roosevelt believed that rural electrification would uplift agricultural regions and provide farmers with a higher standard of living.[71]

During the campaign Roosevelt also embedded his utilities proposals within the underconsumptionist analysis of the Depression. The industrial slump, as he and his advisers ceaselessly claimed, was linked to insufficient demand and low purchasing power. Raising farm income directly with the allotment plan was one part of the solution, but another step toward increasing consumption was to lower industrial prices. Here, as Tugwell argued, public utilities were the worst offenders, for they "refuse to require lowered rates because of declining income, not realizing that the reason for declining income is the refusal to reduce rates." Roosevelt repeated this assertion in his campaign address on power policy in Portland, Oregon, taking aim at the "selfish interests" who failed to see that lower prices would encourage widespread public use. Though Roosevelt made it clear he did not support government ownership of the utility industry, he set down two exceptions to that rule. First, he believed that any community not satisfied with its electric service could set up its own system. Second, he invoked the federal government's sovereignty over the nation's water power resources to propose the public development and distribution of electricity from the Colorado, Tennessee, St. Lawrence, and Columbia Rivers. "Each one of these," he declared, "will forever be a national yardstick to prevent extortion against the public and to encourage the wider use of . . . electricity."[72]

What therefore distinguished Roosevelt from his competitor in the presidential race was not simply the fact that he promised relief for farmers and unemployed workers. Bolstered by a view of the Depression that traced its origins to the lack of consumer buying power, Roosevelt argued that direct assistance and rural resource programs were essential to recovery. Roosevelt had always felt that rural Americans would see their standards of living rise when they used their land properly and when they had the opportunity to purchase electricity at low rates. But he now argued that these things would help farmers become consumers. Unlike Hoover, who did not accept the tenets of the new conservation, Roosevelt regarded rural electrification, soil conservation, and reforestation as measures to lift the Depression, rather than as rewards to be earned only after the economy had recovered. And not only did his farm and power policies reflect this thinking: it also infused one of his most noted campaign promises to put a million men to work restoring marginal and abandoned lands. "It is clear," he explained, justifying this proposal for a conservation corps, "that economic foresight and immediate employment march hand in hand." Roosevelt's position was unmistakable: permanent recovery required the immediate marriage of human and natural resources.[73]

CONCLUSION: TOWARD THE NEW DEAL

Concerned by the steady depopulation of the countryside and by low standards of living in rural areas, two sets of reformers crafted a new conservationist agenda during the 1920s. Electric utility regulators and public power proponents argued that the benefits of interconnection and hydropower development should be shared equitably and distributed at equal cost to rural consumers. Land economists claimed that farm distress was linked to imprudent agricultural expansion onto marginal lands, and that land-use planning might remedy unprofitable operations, market instability, and soil degradation. Animating both the drive to secure electrical current for the farm and the movement to adjust farming to its natural requirements was the desire to bring "balance" and "permanence" to rural areas. These reformers claimed that creating a settled and cultivated rural life required that farmers replace the agricultural systems of the pioneers with more sustainable and profitable operations, and that the benefits of urban centers—modernized homes and alternate employment opportunities—be available to farm families.

The public and political life of the new rural conservation had local, regional, and national dimensions. On the state level, a few Eastern governors attempted to bring power resources under a single regulatory umbrella, requiring that the benefits of long-distance transmission lines and industrial interconnection be shared with sparsely settled and densely populated regions alike. In the upper Midwest, state officials applied the ideas of land economics to the cutover regions, and in the process of doing so set forward an influential model for a national land program. But the new conservation's political center remained in Washington, where utility reformers worked to strengthen federal control over hydropower developments, and where farm advocates demanded a national solution to the deepening agricultural crisis. After 1930, when commodity prices bottomed out and when jobs dried up in the cities, the recommendations of the rural electrification and land utilization movements—the new conservation's two "policy tributaries"— coalesced into a set of emergency and long-range policy possibilities. In the short term, rural recovery demanded an immediate injection of income assistance, quickly followed by the introduction of less wasteful agricultural practices, the retirement of marginal land, and perhaps even the creation of better planned farming communities and subsistence homesteads. In the long term, however, the country needed far fewer, but more productive, farmers. The administrative machinery created to implement the emergency measures might then be transformed into a vehicle for comprehensive agricultural readjustment. Inexpensive rural electricity, meanwhile, would immediately improve farm lives and farm output, while eventually bringing industries and employment opportunities to rural areas.

The new conservation, however, did not develop within an ideological vacuum, but rather emerged as a political counterpoint to the associational character of the 1920s, and to the decade's most committed associationalist— Commerce Secretary and President Herbert Hoover. Though a committed conservationist of the Progressive era variety with a sincere desire to help rural Americans, Hoover did not believe that rural electrification or the farm crisis required "interest-group" solutions or expanded regulatory systems. Convinced to the last that a streamlined industrial system and state-sponsored voluntary and private initiatives would benefit all, Hoover missed an opportunity to capitalize on what by the early 1930s had become a significant new extension of conservationist thought. Not even in the changed context of the Depression, Hoover avowed, would government transmission of power or government-imposed farm prices justify the resulting blows to economic liberty and individual opportunity. Nor would the promise of long-range adjustments justify direct intervention in agricultural markets or the ill-considered sponsorship of new farm colonies.

Hoover's attitude may have been warranted in part by the new conservation's conflicted and opportunistic thinking. It is true that a few planners elegantly reconciled emergency measures to rehabilitate existing farmers and to build new rural communities with their knowledge that the end goal would be a smaller but more prosperous farm population. But what may have been an entirely rational set of proposals was also a clever appeal to many different interest groups and a ready-made justification for a sweeping array of programs. Anyone who favored multipurpose water projects, for example, or the reclamation program, rural electrification, industrial decentralization, subsistence homesteads, marginal land retirement, soil conservation, reforestation, park expansion, or the "back-to-the-land" movement could place his vision somewhere within the initial framework. But how would the government, or even the most dedicated public servants, adjudicate between these often contradictory interests? Was the goal to keep farmers on the land or to encourage them to find employment elsewhere? How exactly would short-term income assistance evolve into long-range conservation adjustments? Such worries, however, were left for the future. In the meantime, Franklin Roosevelt enthusiastically embraced the precepts of the new conservation, and placed rural development policy at the forefront of his strategy to combat the Depression.

NOTES

1. Franklin Roosevelt, "Acres Fit and Unfit: State Planning of Land Use for Industry and Agriculture," address before the Conference of Governors at French Lick, Indiana, June 2, 1931, in Samuel I. Rosenman, ed., *The Public Papers and Addresses of Franklin D. Roosevelt*, vol. 1 (Random House: New York, 1938) (*PPA*), 485–95.

2. Roosevelt to the Congress, January 14, 1935, in Edgar B. Nixon, ed., *Franklin D. Roosevelt and Conservation* (New York: Arno Press, 1972) (*FDRC*), 342.

3. This is the subject of *This Land, This Nation: Conservation, Rural America, and the New Deal* (Cambridge University Press, forthcoming).

4. Ellis W. Hawley has presented the most influential interpretations of American associationalism. See "Herbert Hoover, the Commerce Secretariat, and the Vision of an 'Associative State,' 1921–1928," *Journal of American History* 61 (June 1974): 116–40: *The Great War and the Search for a Modern Order: A History of the American People and Their Institutions, 1917–1933* (New York: St. Martin's Press, 1979); and Hawley, ed., *Herbert Hoover as Secretary of Commerce: Studies in New Era Thought and Practice* (Iowa City: University of Iowa Press, 1981).

5. Annual Message to the Legislature, 1929, PPA, 82.

6. The term "New Conservation" (in capitals) comes from Lewis Mumford, "Regions—To Live In," *Survey Graphic* (*SG*) (May 1925): 152.

7. Gifford Pinchot, "Giant Power," *SG* (March 1924): 561. Holding company statistics are included in the Federal Power Commission's National Power Survey of 1935, *Principal Electrical Utility Systems in the United States* (Power Series No. 2), 10. Treatments of Giant Power include Jean Christie, "Giant Power: A Progressive Proposal of the Nineteen-Twenties," *Pennsylvania Magazine of History and Biography* 96 (October 1972): 480–507; and Leonard DeGraaf, "Corporate Liberalism and Electric Power System Planning in the 1920s," *Business History Review* 64 (Spring 1990): 1–31. Also see Philip J. Funigiello, *Toward a National Power Policy: The New Deal and the Electric Utility Industry, 1933–1941* (Pittsburgh: University of Pittsburgh Press, 1973); D. Clayton Brown, *Electricity for Rural America: The Fight for the REA* (Westport, CT: Greenwood Press, 1980); Jean Christie, *Morris Llewellyn Cooke, Progressive Engineer* (New York: Garland Publishing, 1983); and Thomas P. Hughes, *Networks of Power: Electrification in Western Society, 1880–1930* (Baltimore: Johns Hopkins University Press, 1983).

8. *Report of the Giant Power Board to the Governor of Pennsylvania* (Harrisburg, PA: December 7, 1926); Cooke to Pinchot, October 25, 1923, March 3, 1924 (italics in original), Cooke Papers, General Correspondence, Numerical File, Red Series (NFRS), Box 36, Franklin D. Roosevelt Library, Hyde Park, New York (FDRL).

9. Morris Cooke, "The Long Look Ahead," *SG* (March 1924): 601–04; Cooke, "The Early Days of the Rural Electrification Idea, 1914–1936," *American Political Science Review* 42 (June 1948): 431–47; Cooke to Pinchot, December 12, 1923, December 18, 1923, Cooke Papers, NFRS, Box 36, FDRL; Cooke to Herbert Myrick, December 26, 1923, Cooke to Frank Willits, December 17, 1923, November 11, 1924, Willits to Cooke, January 28, 1924, Cooke Papers, Giant Power Survey Papers (GPSP), Box 191, FDRL; Ralph Blasingame to Cooke, September 16, 1924, November 20, 1924, Cooke Papers, GPSP, Box 192, FDRL.

10. Pinchot, "Interconnection and Service to the Farm," address before the National Electric Light Association, Atlantic City, NJ, May 21, 1924; "Electric Farm Rival of City, Says Pinchot," *North American,* December 9, 1924; Cooke, "The Farmer's Interest In Conservation," no date (ca. 1923–1925), Cooke Papers, GPSP, Box 200, FDRL. Also see Cooke to Pinchot, December 12, 1923, Cooke Papers, NFRS, Box 36, FDRL; and "Pinchot Urges Electric Power as Aid to Farms, Opens Agricultural Products Show With Pledge to Push Development in State," *North American,* January 23, 1924.

11. Pinchot, "Giant Power," *SG* (March 1924): 562; *Report of the Giant Power Board,* 6–7; Cooke, "The Farmer's Interest in Conservation"; Philip P. Wells, "Our Federal Power Policy," *SG* (March 1924): 570; Pinchot to Alfred Smith, March 9, 1923, Cooke Papers, GPSP, Box 206, FDRL; Cooke to Pinchot, March 10, 1924, Cooke Papers, NFRS, Box 36, FDRL.

12. "A League for Superpower," *Literary Digest,* November 13, 1920; William Hard, "Giant Negotiations for Giant Power: An Interview With Herbert Hoover," *SG* (March 1924); "Summary of Statement of Secretary Hoover to the Super Power Conference, New York City," October 13, 1923, "Address by Herbert Hoover to Convention of National Electric Light Association," May 21, 1924, Hoover Papers, Commerce Papers, Box 162, Herbert Hoover Presidential Library, West Branch, IA (HHL); Cooke to Pinchot, October 25, 1923, Cooke Papers, NFRS, Box 36, FDRL; Pinchot, "Interconnection and Service to the Farm"; "Governor Urges Right Citizenship, Warns of Electric Monopoly," *Philadelphia Inquirer,* July 5, 1925; "Pinchot Asks Help of Congressmen, Assails Superpower Plan," *New York Times,* July 20, 1925. Also see DeGraaf, "Corporate Liberalism and Electric Power System Planning" for an excellent comparison of Giant Power and Super Power.

13. Hard, "Giant Negotiations for Giant Power," *SG* (March 1924): 578; Herbert Hoover, "Interconnection is Now Road of Electrical March," *Philadelphia Chamber of Commerce News Bulletin* (June 1925): 11; Hoover, "Why the Public Interest Requires Local Rather Than Federal Regulation of Electric Public Utilities," Address before the National Association of Railroad and Utilities Commissioners, Washington, DC, October 14, 1925; Hoover Papers, Commerce Papers, Box 591, HHL; Hoover to Cooke, April 2, 1924, Cooke Papers, GPSP, Box 207, FDRL. On the Colorado River Compact, see Norris Hundley, Jr., *Water and the West: The Colorado River and the Politics of Water in the American West* (Berkeley: University of California Press, 1975); and Daniel Tyler, "The Silver Fox of the Rockies: Delphus Emory Carpenter and the Colorado River Compact," *New Mexico Historical Review* 73 (January 1998): 25–43.

14. Morris Cooke, "Ontario Hydro-Electric," *New Republic* (June 21, 1922); Adam Beck, "Ontario's Experience," *SG* (March 1924); Pinchot, "Giant Power," *SG* (March 1924) 562; Martha Bensley Bruère, "Following the Hydro," *SG* (March 1924): 594. Also see Robert W. Bruère, "Pandora's Box," in the same issue of *SG.* Cooke corresponded throughout the 1920s with Adam Beck, chairman of Ontario Hydro, and F.A. Gaby, its chief engineer—see Cooke Papers, NFRS, Box 35, FDRL.

15. *Congressional Record,* sixty-eighth Congress, second session, December 8, 1924, 122; "Senator Norris' Reference to Super Power, Grand Island, Nebraska, September 23, 1924," Hoover Papers, Commerce Papers, Box 590, HHL; George W. Norris, *Fighting Liberal* (New York: Macmillan, 1945); Richard Lowitt, *George W. Norris: The Persistence of a Progressive, 1913–1933* (Chicago: University of Illinois Press, 1971).

16. Richard Lowitt, "Ontario Hydro: A 1925 Tempest in an American Teapot," *Canadian Historical Review* 44 (September 1968): 267–74; Lowitt, *George W. Norris,* 197–216, 244–71; Wells to Cooke, February 28, 1923, Cooke Papers, GPSP, Box 207, FDRL; Pinchot to Cooke, April 10, 1924, Cooke Papers, NFRS, Box 36, FDRL; Morris Cooke, "In Re Muscle Shoals, Before the Senate Committee on Agriculture," April 28, 1924, Cooke Papers, NFRS,

Box 36, FDRL. On Muscle Shoals see Norman Wengert, "Antecedents of TVA: The Legislative History of Muscle Shoals," *Agricultural History* 26 (July 1952): 141–47; Judson King, *The Conservation Fight: From Theodore Roosevelt to the Tennessee Valley Authority* (Washington, DC: Public Affairs Press, 1959); and Preston J. Hubbard, *Origins of the TVA: The Muscle Shoals Controversy, 1920–1932* (Nashville: Vanderbilt University Press, 1961).

17. Herbert Hoover, "Muscle Shoals," prepared for President Coolidge, November 6, 1925, Hoover to Dan E. McGugin, November 18, 1925, Hoover Papers, Commerce Papers, Box 422, HHL; Hoover, "A National Policy in Development of Water Resources: Inland Transportation, Irrigation, Reclamation, Flood Control, Power," August 21, 1926, Hoover Papers, Commerce Papers, Box 688, HHL; Hard, "Giant Negotiations for Giant Power," 579; Herbert Hoover, *American Individualism* (Garden City, NY: Doubleday, Page, and Co., 1922), 4, 60.

18. Stuart Chase, "The Tragedy of Waste," *New Republic* 44 (September 1925): 40.

19. Theodore Saloutos and John Hicks, *Agricultural Discontent in the Middle West, 1900–1939* (Madison: University of Wisconsin Press, 1951); James Shideler, *Farm Crisis* (Berkeley: University of California Press, 1957); Donald Worster, *Dust Bowl: The Southern Plains in the 1930s* (New York: Oxford University Press, 1979); David E. Hamilton, *From New Day to New Deal: American Farm Policy From Hoover to Roosevelt, 1928–1933* (Chapel Hill: University of North Carolina Press, 1991); Deborah Fitzgerald *Every Farm a Factory: The Industrial Ideal in American Agriculture* (New Haven: Yale University Press, 2003).

20. Hamilton, *From New Day to New Deal,* 13–21; Gilbert Fite, *American Farmers: The New Minority* (Bloomington: Indiana University Press, 1981), 38–48.

21. Hamilton, *From New Day to New Deal;* Jess Gilbert and Ellen Baker, "Wisconsin Economists and New Deal Agricultural Policy: The Legacy of Progressive Professors," *Wisconsin Magazine of History* 80 (Summer 1997): 280–312. Also see Richard Lowitt, ed., *Journal of a Tamed Bureaucrat: Nils A. Olsen and the BAE, 1922–1935* (Ames: Iowa State University Press, 1980); and Henry C. and Anne D. Taylor, *The Story of Agricultural Economics in the United States, 1840–1932* (Ames: Iowa State University Press, 1952).

22. L.C. Gray, "The Evolution of the Land Program," address before the BAE Conference on Agricultural Planning, March 22, 1939, Loeb Library Vertical File, Harvard University (LVF); Gray, "The Field of Land Utilization," *Journal of Land and Public Utility Economics* 1 (April 1925): 152–59; Gilbert and Baker, "Wisconsin Economists and New Deal Agricultural Policy"; Leonard A. Salter, Jr., *A Critical Review of Research in Land Economics* (Minneapolis: University of Minnesota Press, 1948); Margaret R. Purcell, USDA Division of Land Economics, "A Quarter Century of Land Economics in the Department of Agriculture, 1919–44," October 1945, manuscript courtesy Jess Gilbert, University of Wisconsin.

23. L.C. Gray, "The Evolution of the Land Program," 5; "The Field of Land Utilization"; "The Utilization of Our Lands for Crops, Pasture and Forests,"

Agricultural Yearbook 1923 (Washington, DC: USGPO, 1924): 415–506; Purcell, "A Quarter Century of Land Economics"; Richard S. Kirkendall, "L.C. Gray and the Supply of Agricultural Land," *Agricultural History* 37 (October 1963): 206–14; Albert Z. Guttenberg, "The Land Utilization Movement of the 1920s," *Agricultural History* 50 (July 1976): 477–90; Tim Lehman, *Public Values, Private Lands: Farmland Preservation Policy, 1933–1985* (Chapel Hill: University of North Carolina Press, 1995), 9–17; Melissa G. Wiedenfeld, "The Development of a New Deal Land Policy: Fergus County, Montana, 1900–1945," (Ph.D. diss., Louisiana State University, 1997), 138–67.

24. Gray, "The Field of Land Utilization"; P.S. Lovejoy, "Theory and Practice in Land Classification," *Journal of Land and Public Utility Economics* 1 (April 1925): 160–75; Salter, *A Critical Review of Research*, 20–22, 83–129; Robert J. Gough, "Richard T. Ely and the Development of the Wisconsin Cutover," *Wisconsin Magazine of History* 75 (Autumn 1991): 3–38; J.D. Black and L.C. Gray, *Land Settlement and Colonization in the Great Lakes States*, USDA Bulletin no. 1295 (Washington: USGPO, 1925).

25. Lovejoy, "Theory and Practice"; Vernon Carstensen, *Farms or Forests: Evolution of a State Land Policy for Northern Wisconsin, 1850–1932* (Madison: University of Wisconsin College of Agriculture, 1958), 90–116; George S. Wehrwein, "A Social and Economic Program for Submarginal Agricultural Areas," *Journal of Farm Economics* 13 (April 1931): 276–80.

26. Richard Ely, "Worthless Land: What Can We Do for the Men on It," *The Country Gentleman*, October 25, 1924, quoted in Gough, "Richard T. Ely," 28.

27. Mary W.M. Hargreaves, *Dry Farming in the Northern Great Plains, 1900–1925* (Cambridge: Harvard University Press, 1957); Fitzgerald, *Yeomen No More*, 81–83; Wiedenfeld, "Development of a New Deal Land Policy," 107–37; Milburn L. Wilson oral history, Columbia University Oral History Collection (CUOH); Russell Lord, *The Wallaces of Iowa* (Boston: Houghton Mifflin, 1947), 298.

28. "M.L. Wilson: Thirty Years of Social Engineering," USDA Division of Extension Information, February 16, 1944, M.L. Wilson Papers, Collection 00003, 10:36, Montana State University Archives, Bozeman, Montana; Harry C. McDean, "M.L. Wilson and the Origins of Federal Farm Policy in the Great Plains, 1909–1914," *Montana: The Magazine of Western History* 34 (Autumn 1984): 50–59.

29. M.L. Wilson, "The Fairway Farms Project," *Journal of Land and Public Utility Economics* 2 (April 1926): 156; Harry C. McDean, "Social Scientists and Farm Poverty on the North American Plains, 1933–1940," *Great Plains Quarterly* 3 (Winter 1983): 17–29; and "Federal Farm Policy and the Dust Bowl: The Half-Right Solution," *North Dakota History* 47 (Summer 1980): 20–31; Fitzgerald, *Yeomen No More*, 80–104; Wilson CUOH; Lord, *Wallaces of Iowa*, 300.

30. Gray, "Field of Land Utilization," 137, 138.

31. Gray, "Evolution of the Land Program"; David Weeks, "Scope and Methods of Research in Land Utilization," *Journal of Farm Economics* 11 (October

1929): 597–98; Hugh H. Bennett, *The Soils and Agriculture of the Southern States* (New York: Macmillan, 1921); Douglas Helms, ed., *Readings in the History of the Soil Conservation Service* (SCS Economics and Social Sciences Division, Historical Notes No. 1).

32. Helms, *Readings in the History of Soil Conservation,* 4; Bennett, "Soil Erosion Takes $200,000,000 Yearly from U.S. Farmers," *USDA Yearbook of Agriculture* (Washington: USGPO, 1927), 593; "Soil Conservation: Report of H.H. Bennett to Chief of Bureau," Misc. Papers of Hugh H. Bennett, 1926–1934, National Archives (NA) Record Group 114, Entry 21, Box 2; Hugh H. Bennett, "Soil-Erosion Problem Under Investigation in National Control Program," *USDA Yearbook of Agriculture* (Washington, DC: USGPO, 1932), 344–51. Also see Donald Swain, *Federal Conservation Policy, 1921–1933* (Berkeley: University of California Press, 1963).

33. Ray L. Wilbur and Arthur M. Hyde, *The Hoover Policies* (New York: Scribner's, 1937), 2, 3, 42, 46–47. For an excellent interpretation of Hoover's conservation record, see Kendrick Clements, *Hoover, Conservation, and Consumerism: Engineering the Good Life* (Lawrence: University Press of Kansas, 2000).

34. Bruce Lohof, "Herbert Hoover, Spokesman of Humane Efficiency: The Mississippi Flood of 1927," *American Quarterly* 22 (Fall 1970): 690–700; Clements, *Hoover, Conservation, and Consumerism,* 111–27; Wilbur and Hyde, *The Hoover Policies,* 264–65; Herbert Hoover, "The Improvement of Our Mid-West Waterways," *Annals of the American Academy of Political and Social Science* 135 (January 1928): 15, 22; Hoover, "A National Policy in Development of Water Resources."

35. Hoover, "A National Policy in Development of Water Resources," 6, 19. Also see Wilbur and Hyde, *The Hoover Policies,* 254–86.

36. Harry Slattery to Morris Cooke, October 20, 1927, Cooke Papers, NFRS, Box 49, FDRL; Norris Hundley, Jr., "The West Against Itself: The Colorado River—An Institutional History," in Gary D. Weatherford and F. Lee Brown, eds., *New Courses for the Colorado River* (Albuquerque: University of New Mexico Press, 1986); Linda J. Lear, "Boulder Dam: A Crossroads in Natural Resource Policy," *Journal of the West* (October 1985): 82–94; Jay L. Brigham, "Public Power and Progressivism in the 1920s," (Ph.D. diss., University of California, Riverside, 1992), 126–28; Department of the Interior, Press Releases, October 21, 1929, December 28, 1929, and December 30, 1931, in Hoover Cabinet Papers, Interior, Reclamation, HHL.

37. Lowitt, *George W. Norris,* 454–67; Norris, *Fighting Liberal;* Judson King, "Power Records of Hoover and Roosevelt," (Washington, DC: National Popular Government League, Bulletin No. 157, September 1932); veto message quoted in Thomas K. McCraw, *TVA and the Power Fight, 1933–1939* (Philadelphia: J. B. Lippincott, 1971), 24.

38. "A Comparative Report on the Boulder Canyon and Muscle Shoals Projects," December 1930; "Principles Adopted by the Muscle Shoals Commission on October 29, 1931"; Joseph McMullen to Walter Newton,

December 20, 1930, September 16, 1931, Hoover Papers, President's Subject File, "Muscle Shoals," HHL; King, "Power Records."

39. Herbert Hoover, "Statement Before House Committee on Rivers and Harbors," January 30, 1926, and "The Waterways Outlet From the Middle West," March 9, 1926, Hoover Papers, Commerce Papers, Box 687, HHL; "St Lawrence O.K., Smith Checkmate," *New York Evening Post,* January 3, 1927; Wilbur and Hyde, *The Hoover Policies,* 270–75; *PPA,* 159–206. Also see Daniel R. Fusfeld, *The Economic Thought of Franklin D. Roosevelt and the Origins of the New Deal* (New York: Columbia University Press, 1956).

40. Roosevelt to Hoover, July 9, 1932, Hoover to Roosevelt, July 10, 1932, *PPA,* 204–05.

41. Hamilton, *From New Day to New Deal.*

42. Ibid. Also see Mordecai Ezekiel CUOH.

43. Ibid.

44. Kirkendall, "L.C. Gray"; Hugh H. Bennett, "What To Do With Surplus Land," September 1931, Misc. Papers of Hugh H. Bennett, 1926–1934, NA RG 114, Entry 21, Box 4; L.C. Gray and O.E. Baker, "Land Utilization and the Farm Problem," USDA Misc. Pub. No. 97 (Washington, DC: USGPO, 1930); L.C. Gray, "National Land Policies in Retrospect and Prospect," *Journal of Farm Economics* 13 (April 1931): 231–45.

45. *Proceedings of the National Conference on Land Utilization* (Washington, DC: USGPO, 1932); "Recommendations of the National Conference on Land Utilization," November 23, 1931, Wilson Collection 2100, 23:21; "Land Use Problems Attacked by Two Committees in Four-Day Meet," Hoover Cabinet Papers, Agriculture, Land Use Studies, 1930–1932, HHL.

46. Purcell, "A Quarter Century of Land Economics"; Gray, "Evolution of the Land Program"; Hamilton, *From New Day to New Deal,* 178–80.

47. Hamilton, *From New Day to New Deal,* 180–85; Richard Kirkendall, *Social Scientists and Farm Politics in the Age of Roosevelt* (Columbia: University of Missouri Press, 1966), 20–24. Also see William D. Rowley, *M.L. Wilson and the Campaign for Domestic Allotment* (Lincoln: University of Nebraska Press, 1970).

48. Hamilton, *From New Day to New Deal,* 182–92; Ezekiel to Tugwell, October 20, 1939, Wilson Collection 00003, 7:8; Wilson CUOH. The idea that Wilson and Ezekiel wanted to use the allotment machinery to build "administrative capacities" is elegantly developed in Hamilton, *From New Day to New Deal;* Wilson's description of the allotment plan as a "smokescreen for agricultural planning" comes from a letter he wrote to Ezekiel in April 1932, quoted in Hamilton, 192.

49. M.L. Wilson, "Land Utilization," April 16, 1932, radio address, Economics Series Lecture No. 25, University of Chicago Press; Wilson, "A Land Use Program for the Federal Government," *Journal of Farm Economics* 15 (April 1933): 217–35.

50. Wilson, "Land Utilization"; Wilson, "A Land Use Program for the Federal Government"; Wilson, "A New Land-Use Program: The Place of Subsistence Homesteads," *Journal of Land and Public Utility Economics* 10 (February

1934): 1–12; Wilson, "Planning Agriculture in Relation to Industry," *Rural America* (December 1934): 4; Wilson CUOH; P.K. Whelpton, "The Extent, Character and Future of the New Landward Movement," *Journal of Farm Economics* 15 (January 1933): 57–72; "Back-To-Land Movement Needs Safeguards, National Committee Says," April 8, 1932, Hoover Cabinet Papers, Agriculture, Land Use Studies, 1930–1932, HHL; Gray, "National Land Policies." Also see Kirkendall, "L.C. Gray," and Paul K. Conkin, *Tomorrow a New World: The New Deal Community Program* (Ithaca: Cornell University Press, 1959).

51. Wilson, "Land Utilization"; Hamilton, *From New Deal to New Day,* 195–215; Wilbur and Hyde, *The Hoover Policies,* 152–80; *Public Papers of the Presidents: Herbert Hoover, 1932–33,* 369–70, quoted in Hamilton, 207.

52. "A Hyde Farm Plan," December 30, 1932, and "Statement by Secretary of Agriculture Arthur M. Hyde," February 18, 1933, Hoover Papers, President's Subject File, "Farm Matters—Hyde Farm Plan 1933," HHL; Wilson CUOH.

53. "A Hyde Farm Plan." Documents concerning the southern settlement bills are contained in Hoover Papers, President's Subject File, "Rural Development," HHL. Hoover's reaction is recorded in J.M. Patterson to Hugh McRae, April 25, 1930.

54. Statistics on farm abandonment in New York are included in Bernard Bellush, *Franklin D. Roosevelt as Governor of New York* (New York: Columbia University Press, 1955), 91–92; Address at State College of Agriculture, Cornell University, Ithaca, New York, 1930, *PPA,* 140–43; Annual Message to the Legislature, 1929, *PPA,* 81; Address before the Conference of Governors at French Lick, Indiana, 1931, *PPA,* 485–94. Also see Fusfeld, *Economic Thought of Franklin D. Roosevelt.*

55. "A Proposal for a Survey of Soil and Climatic Conditions," Address at Silver Lake, New York, 1929; *PPA,* 477–79; "A Message to the Legislature Formulating a Land Policy for the State," 1931, *PPA,* 480–85; Annual Message to the Legislature, 1932, *PPA,* 119; Leonard A. Salter, Jr., *A Critical Review of Research in Land Economics* (Minneapolis: University of Minnesota Press, 1948), 130–33; Radio Address on the Conservation of Natural Resources, 1930, *PPA,* 521–25; Radio Address Urging Voters to Support the Reforestation Amendment, 1931, *PPA,* 526–31; Address before the Conference of Governors at French Lick, Indiana, 1931, *PPA,* 485–94. For a critical assessment of New York's land program see Salter, *A Critical Review,* and E. Melanie DuPuis, "In the Name of Nature: Ecology, Marginality, and Rural Land Use Planning During the New Deal," in E. Melanie DuPuis and Peter Vandergeest, eds., *Creating the Countryside: The Politics of Rural and Environmental Discourse* (Philadelphia: Temple University Press, 1996).

56. Address before the Conference of Governors, 487; Address before the American Country Life Conference, Ithaca, NY, 1931, *PPA,* 503–15; Speech by Roosevelt to the Citizens of the State of New York, October 26, 1931, *FDRC,* 98.

57. First Inaugural Address as Governor, 1929, *PPA,* 77–78.

58. Plan for the Development of the Water Power Resources of the St. Lawrence, 1929, *PPA,* 172; Roosevelt, "The Real Meaning of the Power Problem," *Forum* (December 1929), cited in Fusfeld, *Economic Thought,* 139; Campaign Address, Syracuse, NY, 1930, *PPA,* 419–26.

59. *PPA,* 163; Fusfeld, *Economic Thought,* 141; Morris Cooke to Felix Frankfurter, December 19, 1932, Cooke Papers, NFRS, Box 8, FDRL.

60. *PPA,* 163–66; Christie, *Morris Llewellyn Cooke,* 100–05; Cooke to Frankfurter, December 19, 1932; Cooke, "Informal Memorandum on Governor Roosevelt's Power Record," no date (ca. 1932), Cooke Papers, NFRS, Box 51, FDRL; "A Recommendation for Legislation Permitting Municipalities to Build Their Own Power Plants," February 15, 1932, *PPA,* 202–03.

61. Franklin D. Roosevelt, *Looking Forward* (New York: John Day, 1933), 125–36, 147–54.

62. Raymond Moley, *After Seven Years* (New York: Harper & Brothers, 1939), 12–13.

63. Moley, *After Seven Years,* 14–15; Rexford Tugwell, *The Brains Trust* (New York: Viking Press, 1968), 12–22. Also see Gertrude A. Slichter, "Franklin D. Roosevelt and the Farm Problem, 1929–1932," *Mississippi Valley Historical Review* 43 (September 1956): 238–58.

64. Moley, *After Seven Years,* 15; Bernard Sternsher, *Rexford Tugwell and the New Deal* (New Brunswick: Rutgers University Press, 1964), 15–25; Tugwell, "The Responsibilities of Partnership," Address before the Iowa State Bankers Association, June 27, 1934, Address before the Consumers' League of Ohio, May 11, 1934, LVF; Tugwell, *Brains Trust,* 16, 24–25.

65. Sternsher, *Rexford Tugwell and the New Deal,* 26–39; "The Governor Accepts the Nomination for the Presidency," July 1932, *PPA,* 647–59.

66. Tugwell, "Farm Relief and a Permanent Agriculture," *Annals of the American Academy* 142 (March 1929): 271–82; Tugwell, "The Place of Government in a National Land Program," *Journal of Farm Economics* 16 (January 1934): 55–69; Tugwell, *Brains Trust,* 205, 207.

67. Tugwell, *Brains Trust,* 205–08; Michael V. Namorato, ed., *The Diary of Rexford G. Tugwell, 1932–1935* (New York: Greenwood Press, 1992), 42; Richard S. Kirkendall, *Social Scientists and Farm Politics in the Age of Roosevelt* (Columbia: University of Missouri Press, 1966), 42.

68. *Diary of Rexford G. Tugwell,* 42–45; Hamilton, *From New Day to New Deal,* 210–11; Kirkendall, *Social Scientists and Farm Politics,* 41–46.

69. Hamilton, *From New Day to New Deal,* 198–201; Tugwell, *Brains Trust,* 449; Wilson CUOH.

70. Moley, *After Seven Years,* 41–45; Slichter, "Franklin D. Roosevelt"; "Roosevelt's Farm Adviser Tells His Plan," *Chicago Sunday Tribune,* October 2, 1932; Campaign Address on the Farm Problem at Topeka, Kansas, *PPA,* 693–711; Wilson to Ezekiel, October 8, 1932, Wilson Collection 00003, Box 6.

71. Ernest Gruening, "Power as a Campaign Issue," *Current History* 37 (October 1932): 44–50; Judson King, "Power Records of Hoover and Roosevelt" (Washington, DC: National Popular Government League, 1932).

72. *Diary of Rexford G. Tugwell,* 36–37; Campaign Address on Public Utilities and Development of Hydro-Electric Power, Portland, Oregon, September 21, 1932, *PPA,* 727–42.

73. "The Governor Accepts the Nomination for the Presidency," July 2, 1932, *PPA,* 654; "Senator Norris's Appeal to Break 'Party Regularity' for Election of Roosevelt," *New York Times,* October 18, 1932.

PART 3

LAW, POLICY, AND PLANNING

CHAPTER 6

Rediscovering the New Deal's Environmental Legacy

A. Dan Tarlock

I. Introduction: The New Deal as a Blank Space in the History of Environmentalism

THIS BOOK CLEARLY DEMONSTRATES THAT THERE IS MUCH ABOUT THE NEW DEAL THAT CAN BE CHARACTERIZED as environmental once one substitutes the word "environmental" for "conservation." It helps correct a widely held view that the New Deal is virtually a blank space in the history of modern environmentalism. The New Deal carried forward and greatly extended the work of the Progressive Conservation era. But the work represented here and at the stimulating conference from which it was drawn raises a more difficult question: what is the New Deal's legacy for modern environmentalism? This problem becomes all the more acute when one considers that the gap between the social and political conditions and the culture of the New Deal, and the beginning of the twenty-first century seems almost unbridgeable. To be sure, the New Deal was a glorious era, and one can only dream of a president with Franklin D. Roosevelt's (FDR) commitment to the cause of resource conservation and such a deep knowledge of the subject.[1] But we must recognize that the political, scientific, and cultural conditions of America today are very different from what they were during the 1930s and 1940s. Some of the New Deal conservation/environmental experience, such as the fights to dedicate public lands to park and related purposes, continue to resonate today. But many issues in contemporary environmental policy are the consequences of problems that were not as serious then as they are now. Hence, it will be necessary to reinterpret the conservation/environmental policies of the New Deal if we wish to apply them to the complex pollution and biodiversity conservation problems that we face today. This chapter attempts to both recover and reinterpret a legacy that both informs as well as inspires us.

II. THE NEW DEAL IN THE CONTEXT OF ENVIRONMENTAL HISTORY

THE ENVIRONMENTAL DEGRADATION AND REMEDIATION NARRATIVE

Modern environmental history has developed a powerful, depressing narrative to explain the current state of our basic life support systems. The basic story is a series of Gardens of Eden degraded by Western rationalism, colonization and rapid, unrestrained resource exploitation made possible by the Industrial Age. As the consequences became better understood, society gradually awoke to the social costs of viewing nature solely as a commodity for immediate human consumption and profit. The narrative begins with ancient civilizations' loss of their spiritual connection with nature as Greco-Roman rationalism[2] and later Christianity triumphed. Indigenous sustainable resource practices were rapidly swept away by the European colonization of the New World, Africa, and Asia. Awareness of resource abuse awakened in the nineteenth century as technology and the accumulation of capital made it possible to increase the exploitation rates with a concomitant accelerating increase in degradation.[3] The reaction against the consequences of unrestrained exploitation ultimately led to the modern environmental movement. One can trace the narrative back to ancient Mesopotamia and Greece[4] or begin at any point in time. Modern environmental history usually begins with the Romantic reaction to the end product of the Enlightenment, the French Revolution. More recent research has cast doubt on the argument that indigenous peoples were natural stewards of the landscape until Europeans transformed it into exclusive commodities.[5] In short, the historical degradation narrative is clear, convincing and continuing regardless of the human actors or the cultures that molded them.

United States environmental history usually starts with the post–Civil War scientific surveys that led to the public awareness of the adverse consequences of rapid and unchecked regulation of natural resources,[6] largely as a result of the policy of the rapid disposition of public lands. In the late nineteenth century, the fear of resource exhaustion led to a change in public domain policy from almost total disposition to one of limited disposition and increased retention and management.[7] This shift was further enhanced in the Progressive Conservation era (1900–1920) as many progressives equated resource degradation with the sin of waste. Modern environmentalists are often critical of the equation of conservation with wise *use* rather than preservation,[8] but the modern environmental era would not be possible without the earlier one. Conservationists sought to develop a more balanced resource development policy and introduced the idea of federal government stewardship of our natural resources that remains one of the core organizing principles of modern environmentalism.

Government stewardship was asserted primarily to promote the more efficient, long-term use of natural resources. But the Progressive Conservation era contributed at least three crucial, but underappreciated, permanent ideas to our use of natural resources, ideas that form the backbone of environmental regulation: the principle that private property is subject to public interest limitations; that the government should retain and manage public lands; and that all natural resource use should be subject to regulation.[9] This legacy is often minimized by the modern environmental movement,[10] which has chosen instead to focus on conservation's split into the utilitarian "wise use" and preservation movement[11] over watershed management and the construction of Hetch Hetchy dam north of Yosemite National Park. Thus, many environmentalists have missed the continuity between the Progressive era, the New Deal, and the modern environmental movement.

The modern conventional environmental narrative largely stops in 1920 and abruptly resumes in the 1950s.[12] Modern environmentalism is portrayed largely as a post–World War II phenomenon fueled by suburbanization, affluence, concern over "atomic fallout" and other invisible poisons, and a redirection of the social energy directed against the Vietnam War.[13] All these combined to make the idea of resource preservation for aesthetic and spiritual reasons, articulated by John Muir and refined by Aldo Leopold, a mass rather than elite movement. As air and water pollution became more visible and widespread, a local issue was transformed into a national one. Scientists were quick to extend this thinking to the broader idea of ecosystem protection and the need to reduce the gross forms of air and water pollution.

THE NEW DEAL AS A BLANK SPACE IN THE ENVIRONMENTAL NARRATIVE

The New Deal era (1932–1945) is largely a blank space in this narrative.[14] Resource planning and conservation were important parts of the New Deal, but few historians link these efforts to modern environmentalism.[15] In part, this reflects the fact that significant conservation efforts are often subjugated to the grand themes of the period—overcoming the Great Depression and defeating the totalitarian axis in World War II. In this context New Deal resource conservation is generally presented as a response to specific social conditions, such as the westward migration of a significant portion of the Great Plains population due to the drought of the 1930s; as unfulfilled social engineering, as was the case with the establishment of the Tennessee Valley Authority (TVA); as the result of intense but petty high-level political infighting, best illustrated by Interior Secretary Harold Ickes's persistent but unsuccessful efforts to transform the Department of the Interior into a Department of Conservation;[16] or as a noble but largely unsuccessful effort to use rational, national planning to better people's lives.

President Roosevelt's personal commitment to conservation, especially to planting trees, is given its due, but his deep, genuine interest in the health of land is often portrayed as one of the many, somewhat quaint, lessons he learned as a gentleman farmer in Hyde Park,[17] New York. Ironically, many historians see Roosevelt's privileged Hudson River background as something that he *had* to transcend. They rush past Roosevelt's interest in land and forestry to introduce the personalities and trials that shaped his life and propelled him to the presidency. Similarly, his fight against soil erosion,[18] is treated as either a minor chapter in the progression from twentieth-century resource conservation to environmentally sustainable development or as a failed resource management experiment for which a radical new remedy—environmental protection—was necessary.[19] In fact the lessons FDR learned in Hyde Park, including land health and stewardship and the regenerative power of nature, remained central to Roosevelt's thinking throughout his life. More than any other president, he had an acute awareness of the potential limits that the environment places on humans and on the need to understand those limits. Subsequent presidential speech writers have written noble words about conservation but none match the directness and originality of Roosevelt's words in his 1933 Inaugural Address: "we must . . . endeavor to provide a better use of land for those best fitted for the land."[20]

The lack of connection between the New Deal and environmental protection has even been given legal status by at least one state supreme court. In concluding that new federal reserved water rights attached to a wildlife refuge, on an island in Snake River, which were alleged to be inconsistent with the operation of a federal reclamation project, the Idaho Supreme Court opined:

> The reclamation projects . . . assured that there would be sufficient water to maintain the islands without a federal reserved right . . . The only way that this reality fails is if there is a catastrophic drought or other natural disaster that threatens to dry up the river. . . . It is inconceivable that President Roosevelt in 1937, in the context of the dust bowl years intended to give preference to waterfowl, or other migratory birds, over people.[21]

MODERN ENVIRONMENTALISM AS AN OEDIPAL REACTION TO THE NEW DEAL

Modern environmentalism can be explained as an Oedipal reaction to the past generation that appears to reject most, but not all, aspects of the New Deal.

On this reading, the New Deal is more than a blank space; it is *the* source of many modern problems. Modern environmentalism has widely, if not always explicitly, viewed the New Deal as a negative model of public resource management that must be drastically remedied. Modern environmentalism rejected two out of the three foundational ideas of the New Deal and the New Deal view of nature. Although generally associated with the political Left (and the rapidly diminishing remnants of the old Republican Progressive Conservation tradition), one of the many ironies of modern environmentalism is that it rejected the idea that the expert administrative state was the exclusive source of the public interest and that judicial review should play a minimal role in policing the exercise of discretion within the administrative state. However, it enthusiastically built on the New Deal nationalization legacy to federalize responsibility for environmental protection to the maximum extent possible in the name of preventing races to the bottom.

Modern environmentalism decisively rejected the New Deal hubris that nature can and should be improved and managed for human betterment, which was central to the New Deal resource programs such as the Soil Conservation Service and the TVA as well as FDR's extensive tree planting on his expanded Hyde Park estate. Unlike the post-Atomic Bomb and Holocaust generations, New Dealers could still have an almost unlimited faith in science, technology, and rationality to produce an elite, expert administrative state to solve all social and economic problems and to better human kind. Modern environmentalism reflects the post–World War II skepticism of remote "scientific" experts and faith in the administrative state and rational planning processes to define the public interest.

To counter this perceived misplaced hubris, the modern environmental movement used, without any sense of irony, the anti–New Deal idea of the necessity for judicial control of administrative agencies to attack the environmental insensitivity of the mission of public works and resource management programs, many of which were put in place during the New Deal. New Deal administrative law was initially premised on the desirability of minimal judicial control of the exercise of expert discretion. The more conservative idea of subjecting the administrative state to the rule of law ultimately prevailed when the Administrative Procedure Act [APA] was passed in 1946 after a decade of debate. The resistance by many (but not all) influential New Dealers was tempered by the lessons of the Nazis' perversion of law and the legal system.[22] Liberal jurists such as Justice Thurgood Marshall and Judge J. Skelly Wright of the District of Columbia Circuit Court of Appeals used the APA to develop the "hard look" doctrine of judicial review that allowed environmental Non-governmental Organizations (NGOs) to challenge a wide range of resource management and regulatory decisions that had previously been immune from judicial review.[23]

The importance of the APA cannot be overstated. Modern environmentalism would not exist without the APA. Environmental law is in large part an unplanned by-product of the unique politics of environmentalism in the late 1960s and early 1970s. Environmental law began as a legal guerilla movement led by ad hoc groups of citizens who tapped a growing frustration with development and the idea that application of all technology was progress.[24] The objective was often to stop a local public works project or a federal- or state-licensed activity that allowed the development of scenic "natural" areas.[25] Environmental law was born out of the fight to stop a pump storage project at Storm King Mountain on the Hudson River in New York State. The successful lawsuit to remand a Federal Power Commission license became the paradigm environmental lawsuit.[26]

The movement had to begin as a legal guerilla movement because there was little available recourse in the executive or legislative branches except for big-ticket issues such as large federal dams. In the aftermath of the New Deal, conservation remained on the political agenda but the issues centered primarily on the use of western public lands. In the 1960s, lawyers had to create the subject of environmental law from whole cloth and as a result, lawyers followed the great common tradition of marginal groups and pursued a "rule of law" litigation strategy.[27] To discipline public agencies, through what we now call "public interest" litigation, they had to convince courts that environmental law in fact existed when it did not exist. Lawyers had to create the fiction that the recognition of new environmental protection duties *merely* required courts to perform their traditional and constitutionally legitimate function of enforcing, rather than creating, preexisting rules. This strategy was adopted out of necessity in a more or less ad hoc fashion because environmental values had almost no support in the common law, constitutional law, and in legislation. Moreover, access to the courts was limited because standing was thought to be confined to common law rights, statutory rights, or the clear legislative creation of non-common law legal interests.

NGOs convinced courts to create a new environmental law by simultaneously strictly construing statutes to conclude that the agency failed to follow the letter of the statute or by reading statutes creatively to find "environmental duties," duties in broad delegations of discretion. Sometimes a strict statutory reading was fiction to disguise the radical nature of the legal argument; in other cases, lawyers resorted to orginalist theories of statutory interpretation to find environmental protection duties. Occasionally, an agency action was found to be ultra vires on the merits.[28] In short, all environmental litigation proceeded on the premise that judicial intervention was necessary to uphold the rule of law.[29] Lawyers sought to obtain, in Joseph Sax's words, either a legislative or administrative remand.[30] This strategy was adopted by the Supreme Court in *Citizens to Preserve Overton Park v. Volpe*,[31] which imposed a high burden on the Federal Highway Administration to justify the use of parklands for federally funded highways. At the same time that the environmentalists were attacking the planning processes of the public works agencies, they supported the imposing of strict, new planning requirements on the major federal land management agencies such as the Forest Service and the Bureau of Land Management.

III. THE ARGUMENT: THE NEW DEAL HAS IMPORTANT LESSONS TO TEACH

The principal argument in this essay is that the idea of the New Deal as a blank space in modern environmental history or an Oedipal revolt is inaccurate and simplistic because it obscures many valuable lessons and continuities of the period for modern environmentalism, especially as it moves into the second generation. It is this author's view that the roots of many of underlying ideas of modern environmentalism can be found in the New Deal's attitude toward the use of natural resources. To make this argument, it is necessary to look beyond FDR's environmental legacy at the entire New Deal, at the resource management programs that the New Deal launched or extended, at the ideas—even those that were rejected or ignored—of FDR and influential members of his administration and the subsequently important work that was started in the New Deal. This argument does not assert that the history of New Deal should be revised to portray it as a paradigm of environmental sensibility, as the term is currently used. This would be attributing values and constructs that did not and could not exist at the time.

It is not the intention of this author to claim that the New Deal was environmental by today's standards. Nor is it his intention to challenge the main environmental narrative that identifies many of the New Deal programs—such as the subsidization of large-scale public works, the abdication of responsibility for range management to the grazing industry, the failure to develop a sustainable agriculture and the overreliance on expertise—as the cause of modern environmental degradation. Rather, this writer makes the much more modest claim that a closer examination of the New Deal's resource planning and conservation programs reveals that the roots of modern environmentalism can be found in the New Deal's attitude toward the use of natural resources. The New Deal carried forward and expanded much of the Progressive resource conservation legacy, such as National Parks, and it developed a sophisticated, if scientifically naïve, approach to the relationship between humans and nature.

The legacy of the New Deal is of special importance for the future of environmentalism as it moves to confront the hard problems that were glossed over in the heady flush of the 1970s. In the first generation, the environmental movement targeted government agencies and large industries as the primary sources of air and water pollution, which, along with the disposal of hazardous waste, were identified as the major causes of environmental degradation. Ecology was read to mean walling off as much "nature" as possible from sustained human use. This attitude has changed as the focus of environmentalists has shifted to the restoration of degraded ecosystems. After the initial rush of political and legal success in the 1970s, larger questions about the movement's sustainability and integration with the Western legal and philosophical tradition of individual human dignity were largely ignored—including such issues as the relationship between environmental protection and a liberal, free market economy; the responsibility of ordinary citizens (consumers) and of landowners to protect the environment; and whether it is possible to encourage landscapes that both conserve biodiversity and support human use. Deep ecology has been rejected in favor of viewing environmental degradation as a correctable market failure that does not require substantial interference with the established economic order or individual consumption. Hence it is possible to claim one is "green" while driving three blocks in an SUV with a "Save the Whales" bumper sticker to pick up a latte.

The New Deal is particularly instructive because it enjoyed no such immunity from confronting hard questions and political choices. It had to forge its policies in the midst of a social crisis and all its actions were subject to intense, often partisan debates. The economic collapse of industrial and agricultural production that produced the Great Depression forced New Deal thinkers to address the root causes of these problems, including the problem of resource degradation. Above all, the New Dealers had to confront the problem of unsustainable land-use practices, particularly soil erosion—an issue that is slowly coming to the fore of the modern environmental movement as we realize that land-use practices and public land allocation decisions often cancel out the pollution abatement gains from technology. The hard choices and debates of the 1930s—even those that resulted in only theoretical solutions—stand in sharp contrast to the speed with which environmental protection was established as a political priority in the twenty-first century. Here, the *relative* ease with which public and private actors have absorbed the costs of environmental protection have made it easy for them to ignore all the hard questions.

To gain a better understanding of how the New Deal confronted major environmental issues, this essay now turns its attention to four aspects of the New Deal legacy. In Section IV, we examine selected aspects of New Deal National Park policy that served to extend the Progressive era idea of preserving wild lands for aesthetic reasons to the more modern view of a national park as an ecosystem. In this section we also argue that the water pollution polices of the 1930s and 1940s are an important, and now largely forgotten, example of the deep roots of modern pollution control. Section V revisits the Dust Bowl experience and recasts it as an example of sustainable development. Section VI surveys the New Deal's vision of the landscape and offers it as a scientifically obsolete but socially realistic approach to modern conservation and biodiversity problems. Finally, the essay concludes with a suggested list of the most important legacies of the New Deal for modern environmentalism.

IV. THE EXTENSION OF THE PROGRESSIVE LEGACY OF RESOURCE CONSERVATION AND THE ROOTS OF MODERN ENVIRONMENTAL REGULATION

The administration of Theodore Roosevelt (1900–1908) is conventionally credited with making the government retention and management of public lands a national goal.[32] Progressive conservation's contribution to America remains great, but it is important to realize that this legacy was conserved and extended during the New Deal. Without the New Deal's embrace, the Progressive Conservation legacy might be less ingrained in American law, politics, and culture than it now is. One of the ways in which the legacy was extended was the concerted, if not always successful, effort of the resource management agencies to apply state-of-the art science to federal land management issues; the roots of many modern concepts such as the parks as key elements in a larger ecosystem can be found in agency documents.

NATIONAL PARKS

The New Deal's environmental national park record is important but mixed. It made major contributions to the National Park system by extending the park idea from the preservation of scenic wonders to the rudiments of the park as a significant ecosystem. However, it also laid the seeds for many modern controversies by funding programs such as the CCC that increased recreational use of the parks at the expense of ecological management.[33] Modern national park historians often criticize the failure of Congress to create many of the New Deal-era parks as complete biological or ecological units[34] but the New Deal did lay the seeds for the park-as-ecosystem. Wildlife biologists within the National Park Service criticized the Mather era emphasis on tourism because it led to the decline of wildlife. The idea of the greater Yellowstone ecosystem can be found in a 1933 letter from a wildlife biologist to Arno B. Cammerer, who replaced Horace Albright as director.[35]

The creation of the Everglades National Park, which occurred in the late 1930s although the land was not assembled until 1947, is the best example of the creation of a national park around an ecosystem. The park marked the first time that Congress created a park to preserve a primate area for ecological rather than cultural heritage and aesthetic reasons. As Alfred Runte has written, "[t]otally devoid of the mileposts of cultural nationalism, the Everglades confirmed the depth of commitments to protect more than the physical environment" of national parks.[36] President Roosevelt played an important personal role in adding earlier excluded rain forest areas to the Olympic National Park in the late 1930s.[37]

THE NEW DEAL ORIGINS OF THE CLEAN WATER ACT

The story of federal water pollution control is conventionally one of the post–World War II transitions from the entrenched idea that pollution was a state and local responsibility to the current idea that control is a joint federal–state responsibility. Federal control evolved from the late 1930s to the 1970s. Initially, tentative steps to fund pollution control research in the 1940s were followed by a weak federal program to set standards on interstate waters and to provide grants for publicly owned sewage treatment facilities. By 1968, it was clear that it was very difficult if not impossible to link individual discharges to the violation of an interstate standard and that the enforcement process was not working. As the environmental movement gained momentum, Congress quickly adopted the suggestion of sanitary engineers that discharges be controlled at the source through the application of technology. This story is correct, but the roots of modern pollution can be found in the New Deal.

Water pollution was already a serious problem in many industrial states in the 1930s and pollution control was an important part of the New Deal planning efforts to encourage innovative solutions to transborder pollution. A special advisory committee on water flow of the National Resources Committee, chaired by Morris Cooke, identified water pollution as serious national problem.[38] Water pollution control legislation was introduced in Congress between 1935 and 1944, but no bill was passed. Water pollution did not have the immediacy of other resource degradation problems that the New Deal faced, and thus the case for a federal solution was more controversial. The question of federal versus state responsibility could not be resolved, but bills contained many of the elements of future policy such as a federal inventory of water quality, federal loans or grants for pollution-control facilities, and federal enforcement.

The New Deal had more success at encouraging regional solutions to pollution problems, an approach that remained in vogue through the early 1970s until the Clean Water Act adopted national technology-forcing standards that continue to keep water pollution on the political agenda. The thrust of New Deal planning in all resource areas was to encourage solutions that were appropriate to the geographic scale of the problem. Pollution was largely viewed as a local issue, but the New Deal helped focus attention on its regional nature. The establishment of the National Resources Board, for example, led to the recommendation that states use the interstate compact to address regional problems such as pollution control. Compacts were negotiated on the Delaware, the Potomac, and most notably the Ohio, with the Ohio compact being perhaps the first major water pollution success story prior to the passage of the Clean Water Act.[39]

V. THE DUST BOWL: A CASE STUDY IN THE ORIGINS OF ENVIRONMENTALLY SUSTAINABLE DEVELOPMENT

The Roosevelt administration's response to the Dust Bowl is often cited as the most important New Deal conservation initiative. But, the more interesting story, as Donald Worster has written, is about how the New Deal approached the problem rather than what they did in the end. In contrast to the way that we have addressed many current environmental problems, many New Dealers tried to identify the root causes of resource degradation and to address them directly. Too often modern environmentalism tends to suggest incremental mitigation strategies that eliminate some form of environmental degradation but in the end do not arrest the overall problem. We would now call what many New Dealers tried to accomplish—"sustainable development." It is almost impossible from a perspective of over 60 years to imagine the human and environmental tragedy of the drought and high winds of the 1930s. Morris Cooke's 1936 *Report of the Great Plains Drought Area Committee* starkly sets out the statistics such as the decline in wheat yields in a Kansas county from 16 bushels per acre in 1930 to 2 in 1934. What is striking today is the message that the Report delivered. Following John Wesley Powell, the committee stated directly that humans must adapt to the harsh, unforgiving climate of the Great Plains, not vice versa. Both land- and water-use practices had to change. "The basic cause of the present Great Plains situation is an attempt to impose upon the region a system of agriculture to which the Plains are not adapted to bring into semi and arid region methods which, on the whole, are suitable only for a humid region."

This bold diagnosis was too far ahead of its time. Rather than address the root problem, we implemented incremental measures. In the end more modest, incremental responses were implemented that created the current dysfunctional rural welfare economy that is sustained only by the subsidies that our irrational federal system produces. New farming methods such as fallowing, greater crop diversity, contouring, and stubble retention were put in place along with the famous shelterbelts. Almost seventy years later, the bill for this timidity has come due. The depopulation of rural areas, especially the northern Great Plains, is now a major social issue[40] but, unlike the New Deal, it is not a political one.

The scientific response to the Dust Bowl is important because it was one of the first systematic attempts to apply ecology to a resource degradation problem. Influential ecologists such as Frederick Clements and Paul Sears, whose later book *The Subversive Science* would have a profound impact on the environmental movement, tried to apply the then-dominant succession-climax theory to the Great Plains. Ecologists used powerful explanations of how man had disturbed the plains and argued that a "more complete ecological approach"[41] would permit farmers to exploit the plains without destroying them. In his book, *Dust Bowl,* Donald Worster faults ecologists for not linking the notion of natural, inherent limits to resource exploitation with the need to challenge the fundamental tenants of the Western liberal tradition of capitalism.[42] However, this is too harsh a criticism. Many New Dealers tried to get to the fundamental cause of resource degradation. The *Report of the Great Plains Drought Area Committee,* which attempted to develop a land-use policy for the Great Plains based on the inherent limitations of the region for intensive agriculture and human settlement, represents a genuine, early example of a push toward environmentally sustainable development. It failed partly because the New Deal could not resolve the fundamental dilemma of modern agriculture. The New Deal both tried to restrain and induce production. In the end, then as now, the political forces supporting subsidized production prevailed over this rather remarkable early attempt to make harsh decisions about how much to produce and where.

VI. THE NEW DEAL VISION OF THE LANDSCAPE

The deepest lesson that the New Deal has to teach is the need to have landscape vision that it both grounded in science and social equity. Modern environmentalism has a well-defined, if seldom articulated, vision of the earth's landscape. It is a wild nineteenth century romantic landscape virtually devoid of human beings but with flourishing threatened and endangered species. Most people realize that this landscape is nonexistent in the United States except in a few isolated, protected areas and that it is not a model for a comprehensive environmental policy. It excludes the city from any positive role in environmental protection and more generally it rejects the idea that humans can coexist with the maintenance of important ecosystem services.

The New Deal's landscapes were real places where people had abused the land to their own economic and social detriment. New Deal planners and students of natural resources devoted a great deal of attention to land- and water-use planning and developed a more sophisticated view of the landscape and the processes by which it should be managed. As Daniel Botkin and others have argued, the scientific basis of New Deal policy—the equilibrium paradigm in ecology—has been discredited and replaced by more complex, dynamic visions of the landscape.[43] This landscape is not the one that many modern environmentalists would like to see, but the New Deal dealt with the nation's public and private land in a comprehensive and clearheaded way that provides useful lessons for second-generation environmentalism. Second-generation environmentalists, with some strong dissent,[44] are moving toward a more integrated view of landscape, a view that recognizes humans as part of functioning ecosystems and human use as an integral part of any proposed protection regime.[45] The New Deal accepted this premise without question. In the 1930s, Aldo Leopold had not yet crystallized his principle "let nature be." Hence, the focus of the New Deal was to restore the landscape for human betterment,[46] even though President Roosevelt's 1935 *Message to Congress on Natural Resources* indicated that he understood the need to blend human betterment with respect for "Nature's immutable laws," as his famous phrase, "throwing out of balance of the resources of Nature throws out of balance the lives of men," indicates.

President Roosevelt was committed to the idea of national resource planning to produce more coordinated, efficient new dams, flood-control structures, and other public works. In 1937, the president recommended the creation of seven national planning regions, some organized on river basin lines and others such as the Great Plains on the existence of common environmental problems.[47] The most interesting aspect of this proposal from a modern perspective is that the proposed planning process was a nested one. Recommendations for integrated land and water resource plans were to filter from the local to the state to the regional and then national level. Congress would have the final decision about which projects were to be completed, but the hope was that these plans would provide "a complete picture not only of the needs of each one of the regions but the relationship of each of the regions to the whole of the nation." [48]

The planning dream was not to be. New Deal natural resources planning efforts were hampered by the widespread perception that the New Dealers, especially Rexford Tugwell, were trying to introduce Soviet-style planning programs into the United States.[49] In 1933, for example, Secretary Ickes created the National Planning Board in the Department of the Interior.[50] The Board concentrated on planning for the many proposed river basin development projects and on the creation of supporting state boards. Eventually, the Board morphed into National Resources Committee in 1935. The Committee employed an extremely talented group of experts such as the distinguished geographer and student of water resources policy, Gilbert White. At the same time, an ambitious grassroots planning effort was being developed in the Department of Agriculture as part of land reform and rural resettlement program, which, under the auspices of the Farm Security Administration, tried for a period of ten years to convince farmers to embrace the planning process.[51]

New Deal planning has been harshly criticized as a failed experiment. It was easy to plan short-term projects, but much harder to implement new ideas or establish long-term plans for the future.[52] Nevertheless, the New Deal tried to take a more integrated view of the landscape by linking such elements as rivers and their corridors. Committee reports stressed the need to view the landscape as an integrated whole.[53] New Deal planners urged that as much attention be given to flood plain management as to project development. Unfortunately, this vision was never realized as by the 1940s water resources planning increasingly detached rivers from their watersheds and landscapes. Still, we should recognize that it is the early New Deal vision that animates current river and watershed restoration efforts.[54] Moreover, in the New Deal, process was never divorced from substance and national responsibility.

The New Towns program carried out by the Resettlement administration represents another often overlooked example of the New Deal's efforts to improve the landscape through a more sensitive human footprint that addressed pressing social needs. This Department of Agriculture agency was created after the Supreme Court declared the Industrial Recovery Act unconstitutional.[55] The New Deal New Towns program had its origins in the English Garden City movement. The efforts to adapt this to the American environment were most notable in the planned community of Radburn, New Jersey.[56] The Garden City idea, with its separation of the automobile from pedestrian traffic and the use of cluster development for useable open spaces, was seen as out of touch with the American desire for the traditional suburb of detached single family houses extending ever outward from the old central cities that had developed in the nineteenth century and dominated the twentieth.[57] As we have seen, the 1930s were a very creative period in the history of urban planning and as was the case with rural America a number of major figures such as Frank Lloyd Wright and Lewis Munford tried to come to grips with the emergence of industrial urban America.[58]

Under the guidance of Rexford Tugwell and Stewart Chase, the Towns Program tried to link job creation and affordable housing with the creation of a compact, urban environment that would improve the urban social environment at affordable prices. Leading planners were engaged in the process and the design, and with the exception of Greendale outside of Milwaukee, built low-rise housing blocks "faintly expressive of early 20th century modernism."[59] For many years, the New Deal New Towns program was seen as one of the many failed New Deal utopian social experiments. Only three communities were actually built—outside Washington DC, Cincinnati, and Milwaukee. Federal responsibility shifted from agency to agency as the New Deal state evolved, and the communities were eventually privatized in the Housing Act of 1949 and sold to private developers in the 1950s. Today, they are modestly distinctive islands in the midst of a sea of bland post–World War II suburban development. They did not mark the beginning of either direct federal control of land in the United States or a major federal planning effort.[60] In addition to these programs, historic preservation was also actively promoted during the New Deal and this effort laid the foundation for the post–World War II boom in historic preservation and, along with the writings of such urbanists as Jane Jacobs, for the growing preference for urban as opposed to suburban living.

Time is proving the worth of this New Deal experiment, for the search for an environmentally sensitive urban form[61] has been one of the persistent themes of city planning. In recent years, the old New Town idea has been reborn as the New Urbanism,[62] an international land-use planning and development reform movement that seeks to recreate the compact European city by reconnecting people to a sense of place in contrast to the disbursed post–World War II suburban development pattern. Unfortunately, many new urbanism communities, like the planned unit developments that invoked the Garden City, are essentially gated middle-class redoubts rather than the bold social experiments envisioned by Tugwell. In addition to relieving a severe housing shortage and providing unemployment relief, Tugwell had intended that the residents to grow their own food. During the mid-1930s, some two million people had left cities for farms to survive. As one of the architects of New Deal agricultural policy, Tugwell was well aware that the United States had too many farmers and argued strenuously that the New Deal should separate a portion of food production from farming. Encouraging urban residents to grow a good share of their food was one answer to this crisis—an idea, like New Towns themselves, that has been reborn.

The current effort to address pollution and biodiversity conservation problems through watershed planning is another example of a process that could benefit from revisiting the New Deal's approach to water issues. The lessons are not easy to translate as New Deal planning was directed to building activities that were complimentary to land use whereas today's planning is directed more at changing land-use practices and developing new regimes to manage an existing infrastructure. But, the deeper lesson of a holistic, positive vision of the landscape that works is one that can serve the environmental movement well.

In addition to specific initiatives, the New Deal was a time of fertile intellectual activity, and many of the leading environmental thinkers developed ideas and skills during this period that would shape the modern environmental movement. Rachael Carson is one example. Ms. Carson worked for the United States Fish and Wildlife Service from 1936 to 1952 where she developed her powerful theory—fully articulated in *Silent Spring*—that humans and nature are interdependent but that humans have the power to cause irreversible damage to nature.[63] The aesthetic origins of modern environmentalism have all but been lost as the movement has become dominated by science and economics. But, initially images of nature helped make millions of people aware of the need for environmental protection. Ansel Adams's photographs played a powerful role in the modern environmental movement by introducing people to the beauty of nature and subsequently the need to preserve it.[64] Adams was a leading exponent of documentary or photographic realism that flowered during the New Deal. His realistic photographs were actually elaborately constructed images designed to convey a personal vision. In the 1950s and 1960s, his photographs became concrete manifestations of the objects of environmental protection.

VII. CONCLUSION

In conclusion, this author would argue that the New Deal has at least five ideas that can inform modern environmentalism. The first is that humans and nature are part of a single system, that we are an integral part of the landscape but that human development must be sensitive to natural processes. The efforts of New Deal planners to integrate better land-use planning into programs such as flood control and the Civilian Conservation Corps and National Park projects are examples of how humans might live in closer harmony with natural systems. The second is the core idea of modern pollution control and environmental quality, once again contested, that pollution control and environmental quality is a federal responsibility because of its interstate character. The third is the importance of an overarching idea that resonates with large segments of the population. We have promoted the concept of "biodiversity conservation," but the strategy does not resonate well because it strikes no emotive chord among the public such as respect, care, or love.[65] In contrast, President Roosevelt's idea of resource stewardship, while not as advanced as Aldo Leopold's idea of the land ethic, illustrates the extent to which the New Deal tried to curb excessive private resource exploitation and carried forward and extended the Progressive era conservation movement with a simple but powerful idea that connected stewardship and conservation to individual human actions. Indeed, Roosevelt's personal and public idea of stewardship combined with the Progressive notion that there was a social interest component to private property is a more operational concept than Leopold's land ethic, which relies almost exclusively on individual internalization of the norm. Fourth, modern environmentalism has a bipolar view of environmental protection. It both relies on central command and control and rational planning as well as on more direct democratic involvement in resource decisions. A good share of environmental litigation is directed at expanding the planning horizon of New Deal mandates. But the idea that resource use should be based on sound planning and scientific information has its roots in the New Deal, though this is often ignored in the attack of the specific outcomes of the process. The National Environmental Policy Act was explicitly modeled on New Deal planning. Finally, a better appreciation of the New Deal legacy helps illuminate many of the contradictions of the modern environmental movement and may help it make the painful transition from a largely negative to an affirmative political movement.

NOTES

1. In his book on New Deal national park and related public land policy, Irving Brant recounts that FDR took a trip to the Pacific Northwest at the time that enlargement of the recently created Olympic National Park was being opposed by the U.S. Forest Service. FDR closely questioned the Regional Forester's economic case against the creation of the National Park, factually refuted the misleading data and subsequently (but unsuccessfully) ordered Secretary of Agriculture Henry Wallace to remove the official from the Portland office. *Adventures in Conservation with Franklin D. Roosevelt* 88–89 (Northland Publishers, 1989).

2. J. Donald Hughes, *Pan's Travail Environmental Problems of the Ancient Greeks and Romans* (Baltimore: Johns Hopkins University Press, 1994).

3. See John R. McNeill, *Something Under the Sun: An Environmental History of the Twentieth Century World* (New York: W.W. Norton, 2000).

4. See J. Donald Hughes, *An Environmental History of the World: Humankind's Changing Role in the Community of Life* (London: Routledge, 2001) for a recent synthesis of this narrative.

5. For instance, Shepard H. Ketch, *The Ecological Indian: Myth and History* (W.W. Norton & Co, 2000) and Tim Flannery, *The Eternal Frontier: An Eco-History of North America and Its Peoples* (Grove Press, 2002).

6. William Goetzman, *Exploration and Empire* 356 (Texas State Historical Association, 1966).

7. Samuel P. Hays, *Conservation and the Gospel of Efficiency: The Progressive Conservation Movement 1890–1920* (Boston: Harvard University Press, 1959). But compare Leigh Raymond and Sally K. Fairfax, Fragmentation of Public Domain Law and Policy: An Alternative to the "Shift-to-Retention" Thesis, 39 *Natural Resources Journal* 649 (1999).

8. Ted Steinberg, *Down to Earth: Nature's Role in American History* (Oxford University Press, 2002), 138–56 is highly critical of the conservation movement's forest and park policies for their environmental consequences and cultural deception.

9. See John R. Elliot, T.R.'s Roosevelt Country: Wilderness Legacy, 162 *National Geographic* 340 (September 1982).

10. For a modern critique of the failure to appreciate this legacy, see Peter Huber, *Hard Green: Saving the Environment from Environmentalists* (Philadelphia: Basic Books, 2000).

11. Gifford Pinchot, the second head of the Forest Service, is usually considered the great opponent of preservation for preservation's sake because of successful effort to construct Hetch Hetchy dam and reservoir north of Yosemite National Park. For a good account of the Hetch Hetchy controversy see Richard White, *"Its Your Misfortune and None of My Own": A New History of the American West* 412–15 (Norman Oklahoma: University of Oklahoma Press, 1991). However, his most recent biography, Char Miller, *Gifford Pinchot and the Making of Modern Environmentalism* (Island Press, Washington, DC, 2001), attempts to show that the conventional narrative, e.g. Richard White, supra, is too simplistic. But cf. Down to Earth, supra note 8 at 141, which endorses the conventional view that Pinchot "elevated economics over ecology."

12. For example, in a recent article, the leading U.S. environmental law scholar, Joseph Sax, describes the three eras of environmental protection as: (1) the walling off of nature that occurred in the Progressive Era (2) the control of air, water, and soil pollution that began in 1970, and (3) the modern era of ad hoc, place-specific biodiversity conservation deals. Joseph L. Sax, Environmental Law at the Turn of the Century: A Reportorial Fragment of Contemporary History 88 *California Law Review* 2375 (2000).

13. Samuel P. Hays (with Barbara Hays), *Beauty, Health and Permanence: Environmental Politics in the United States, 1955–1985* (Cambridge University Press, 1987).

14. Frank Graham, Jr. listed the significant landmarks in the history of environmental protection from the perspective of the Audubon Society starting in 1899. There are 21 entries for the period 1899–1920 and 13 for the New Deal. http://magazine.audubonorg/century.html *Down to Earth*, supra note 8 is a prime example. The only mention of the New Deal is to indict it as an early cause of post–World War II suburban sprawl by "funneling huge amounts of money into building roads but little into mass transit." Idem. At 212.

15. Richard N.L. Andrews, *Managing the Environment: Managing Ourselves* 161–78 (New Haven and London: Yale University Press, 1999) is an exception. See also John Opie, *Nature's Nation: An Environmental History of the United States* (New York: Harcourt Brace College Publications, 1998).

16. See T.H. Watkins, *Righteous Pilgrim: The Life and Times of Harold L. Ickes 1874–1952* Part VII (New York: Henry Holt and Company, 1990). Arthur M. Schlisgner, Jr., *The Coming of the New Deal* 343–51 (1958).

17. For instance, Frank Friedel, *Franklin D. Roosevelt: A Rendezvous With Destiny* 6 (Boston, MA: Little, Brown and Co., 1990). Kenneth S. Davis, *FDR: The Beckoning of Destiny 1882–1928* 78–79 (1971) details Roosevelt's interest in ornithology but infers that its only lasting legacy was that it gave him sufficient social poise to overcome his innate shyness.

18. Note 5 at 265–66. The leading work is Donald Worster, *The Dust Bowl: The Southern Plains in the 1930s* 7–8 (London: Oxford University Press, 1979). Other historians generally follow Worster's analysis that the New Deal failed to resolve the tension between the limited carrying capacity of the Great Plains and the desire to people it with family farms. Idem at 155. E.g., Richard White, *It's Your Misfortune and None of My Own*, supra note 11 at 475–81.

19. See Worster, supra note 18 at 229–30 and sources cited notes 11 to 18, *infra*.

20. Inaugural Address, March 4, 1933, 2 *The Public Papers and Addresses of Franklin D. Roosevelt With a Special Introduction and Explanatory Notes By President Roosevelt* 13 (New York: Random House, 1938). [Hereinafter cited as The Public Papers]. See also Excerpts From a Press Conference on the Proper Use of Land, January 24, 1936, 5 *The Public Papers* (1938) at 66 and "We Seek to Pass on to Our Children a Richer Land- a Stronger Nation," Address at Dedication of Shenandoah National Park, July 13, 1936, Idem at 238–40.

21. United States v. Idaho, 23 P.3d 128–29 (Idaho 2001).

22. George D. Shepard, *Fierce Compromise: The Administrative Emerges from New Deal Politics*, 90 Northwestern Law Review 1557 (1996).

23. See notes 21 to 22, *infra*.

24. Former Secretary of the Interior Stewart Udall describes, Victor Yannacone, the first lawyer to try and stop the use of DDT, as follows: "Yannacone was a brilliant tactician, but from the beginning he had no illusions that litigation would produce resounding victories. His maverick motto was 'Sue the Bastards,' and he envisioned his lawsuits as show trials to dramatize environmental truths that would ultimately compel members of the legislative and executive branches of government to act. He was willing to lose court decisions if his cause prevailed in the court of public opinion." *The Quiet Crisis and the Next Generation* 224 (Gibbs Smith Salt Lake City, 1988).

25. In his history of the modern environmental movement, Samuel P. Hays stresses the grass roots, bottoms up nature of the movement compared to the top down elite scientific conservation movement. *Beauty, Health and Permanence* (1987).

26. Scenic Hudson Preservation Conference v. FPC, 354 F.2d 608 (2d Cir. 1965), *cert. denied*, 384 U.S. 941 (1996). The NGO is now Scenic Hudson, Inc., which in 2002 purchased land near the Roosevelt Library to prevent further strip development on Route 9.

27. This discussion is drawn from 2000 William Lloyd Garrison lecture at the Pace University School of Law. A. Dan Tarlock, The Future of Environmental "Rule of Law" Litigation, 19 *Pace Environmental Law Review* 575 (2002).

28. See notes 25 to 27, *infra*.

29. The classic articulation of this argument is David Sive, Some Thoughts of an Environmental Lawyer in the Wilderness of Administrative Law 70 *Columbia Law Review* 612 (1970). Now Professor Sive argued that courts must intensify judicial review of decisions that failed to give adequate consideration to environmental values in order to create "a body of doctrines that will enable

its practitioners to win cases, to neutralize the effluent of affluence, and to prevent the asphalt jungle from supplanting the still green part of our one earth." Idem at 614. This is a restatement of the A.V. Dicey's primary rationale for the rule the law—the necessity to curb arbitrary state power.

30. Joseph L. Sax, *Defending the Environment: A Strategy for Citizen Action* 175–92 (New York: Knopf, 1971).

31. 401 U.S. 402 (1971).

32. Edmund Morris, *Theodore Rex* (New York: Random House, 2001).

33. Richard West Sellars, *Preserving Nature in the National Parks: A History* 132–47 (New Haven: Yale University Press, 1997). A 1991 collection of essays on the Greater Yellowstone Ecosystem Idea, *The Greater Yellowstone Ecosystem: Redefining America's Wilderness Heritage*, Robert B. Keiter and Mark S. Boyce eds. (New Haven and London: Yale University Press, 1991) lays the blame for the subordination of wildlife and ecosystem management to tourism to the Mather-Albright era and the New Deal and dates the ecosystem idea from the 1960s and 1970s. E.g. John J. Craighead, *Yellowstone in Transition*. Idem at 27.

34. Alfred Runte, *National Parks: The American Experience* 138–54 (Lincoln, Nebraska: University of Nebraska Press 2nd ed. revised 1987).

35. Sellars, *Preserving Nature in the National Parks*, supra note 33 at 92.

36. Idem at 137.

37. Brant supra note 1 at 85–144.

38. See N. William Hines, Nor Any Drop to Drink: Federal Regulation of Water Quality Part III: The Federal Effort, 52 *Iowa Law Review* 799, 805–08 (1967) for a brief history of New Deal efforts to pass pollution control legislation. For an analysis of the Water Committee's contribution to national water resources policy see *New Deal Planning* note 52 at 111–99.

39. Edward J. Cleary, *The ORSANCO Story: Water Quality Management in the Ohio Valley under an Interstate Compact* (Baltimore: Johns Hopkins Press, 1967).

40. The United States Bureau of the Census defines a frontier county as one with less than seven persons per square mile. Montana has 47 such counties, South Dakota 39, and North Dakota 37. The respective populations of these states in frontier status are 35, 25, and 24. Kansas and Nebraska also have large numbers of frontier counties but much less of the population lives in these areas due to the urban areas being in the semi-humid eastern areas of these states. The concept of frontier is being defined by professionals to try and better understand the diversity of rural areas, but the continued rural to urban migration in the northern Great Plains, high drug use in rural areas, the inability to attract recreation-oriented in-migration in comparison to the Inner-Mountain West and the increasing poverty rates all point to a need to recapture the New Deal's focus on these areas.

41. H.L. Walster, "Backgrounds of Economic Distress in the Great Plains," in John Hoyt, *The Drought of 1936*, U.S. Geological Supply Paper No. 820, 1938, quoted in Worster, *Dust Bowl* at 201.

42. *Dust Bowl*, supra note 18 at 206–09.

43. See Fred Bosselman, *What Lawmakers Can Learn From Large-Scale Ecology*, 17 Journal Land Use & Environmental Law 207 (2002).

44. See Oliver Houck, Are Humans Part of Ecosystems 28 *Environmental Law* 1 (1996)(humans are part of ecosystems but not the measure of them).

45. See *Panarchy: Understanding Transformations in Human and Natural Systems* (Lance H. Gunderson and C.S. Holling 2002).

46. See Harold Ickes, "Saving the Good Earth: The Mississippi Valley Committee and Its Plan," *Survey Graphic* February, 1934, vol. 23, no. 2.

47. "The President Recommends Legislation for National Planning and Development of Natural Resources Through Seven Regional Authorities," 5 *The Public Papers* at 252.

48. Idem at 255.

49. In 1976, the Indiana House of Representatives Interim Study Committee to Study Federal Regionalism heard testimony from Colonel Arch C. Roberts that federal regionalism was an unconstitional socialist conspiracy. Colonel Roberts invoked that "old revolutionary" Regford Tugwell's book, *The Newstates Constitution*, as an example of a blueprint to give "the appearance of legality to criminal actions of planners and the Congress which precipitated the conspiracy." www.webacessnet/commic/SocAmer.html.

50. Mel Scott, *American City Planning* 301 (Berkeley: University of California Press, 1971).

51. See Richard S. Kirkendall, *Social Scientists and Farm Policies in the Age of Roosevelt* (Columbia Missouri: University of Missouri Press, 1966).

52. Marion Clawson, *New Deal Planning* (Baltimore: Johns Hopkins University Press, 1981). This assessment is repeated in Alan Brinkley, *The End of Reform: New Deal Liberalism in Recession and War* (New York: Knopf, 1995).

53. National Resources Committee, *Regional Factors in National Planning* (Washington, U.S. GPO, 1935).

54. Rural planning was swept away by the resurgence of conservatism in the 1940s and by the New Deal and subsequent administration's embrace of large-scale consolidated farming. See Jess Gilbert and Alice O'Connor, *Leaving the Land Behind: Struggles for Land Reform in U.S. Federal Policy. 1933–1965*, Land Tenure Center University of Wisconsin, Madison Paper 156 (July 1996).

55. *American City Planning* at 336.

56. Radbun had the misfortune to open in May 1929 and the developer went bankrupt in the Great Depression. See C.S. Stein, *Toward New Towns for America* 37 (Cambridge MA: MIT Press, 1957).

57. Kenneth Jackson, *Crab Grass Frontier* (Oxford: Oxford University Press, 1985).

58. Among the notable (now classic) urban planning books published during the New Deal, Harland Bartholomew, *Urban Land Uses* (Cambridge MA: Harvard University Press, 1932), Stuart Chase, *A New Deal* (New York: Macmillan, 1932), Alfred Bettman, *City Planning in Depression and Recovery* (Cambridge MA: Harvard University Press, 1933), Le Corbusier, *The Radiant City* (New York: Orion Press, 1967), Catherine Bauer, *Modern Housing*

(Boston: Houghton Mifflin, 1934), Frank Lloyd Wright, *Broadacre City* (1935), Edward M. Bassett, *The Master Plan* (New York: Rusel Sage, 1938) and Lewis Mumford, *The Culture of Cities* (New York: Harcourt Brace, 1938).

59. *New Deal Regional Planning and Greenbelt Towns,* www.arch.columbia. edu/Projects/Faculty/Call-it-home/Chapter3-nar.html. In an informal talk at Greenbelt, Maryland in 1936, President Roosevelt described the town as "an achievement that ought to be copied in every city in the nation." 5 *The Public Papers* at 599.

60. After the demise of the Natural Resources Planning Board in 1943, the most sustained effort to develop a national land-use planning system occurred in 1970 when Senator Henry Jackson of Washington State introduced legislation to create a cooperate federal, state, and local planning legislation. See Jerold S. Kayden, National Land Use Planning in America: Something Whose Time Has Never Come 3 Washington University J. Of Law & Policy 415 (2000).

61. The leading works are Ian L. McHarg, *Design With Nature* (New York: Doubleday, 1971) and Kevin Lynch, *A Theory of Good City Form* (Cambridge MA: MIT Press, 1981).

62. Information can be found at www.newurbanism.org.

63. See Linda Lear, *Rachael Carson: Witness for Nature* (New York: Henry Holt & Co., 1997).

64. Peter Barr, *Ansel Adams, America's Saint George of Conservation,* Housatonic Museum of ART, http://www.hctc.commnet.edu/artmuseum/anseladams/barressay.html, points out that Adams first developed his style about pure love for the landscapes that he photographed and then became a political activist when he supported the creation of Kings Canyon National Park. Atlantic Monthly article.

65. See Holly Doremus, Biodiversity and the Challenge of Saving the Ordinary 38 *Idaho Law Review* 325 (2002).

CHAPTER 7

FDR's Expansion of Our National Patrimony: A Model for Leadership

JOHN LESHY

IN VIEW OF THE RICH DISCUSSIONS WE'VE HAD SO FAR, I WANT TO BEGIN BY GIVING ONE VERY CONCRETE ILLUSTRATION OF THE IMPACT OF FRANKLIN DELANO ROOSEVELT (FDR's) ENVIRONMENTAL LEGACY on contemporary decision-making. As most of you know, President Clinton energetically employed the Antiquities Act to create new national monuments—protected areas—on federal lands during his term of office. It was a highlight of my public service to help craft these with Secretary of the Interior Bruce Babbitt and the White House.

The Antiquities Act, enacted in 1906 when Teddy Roosevelt (TR) was in the White House, has had a long and honorable history. Every president in the twentieth century except Nixon, Reagan, and the two Bushes used it to protect worthy areas of federal land. But the Act is most prominently associated with TR, who set the tone for its expansive use by protecting the Grand Canyon and some other wonderful areas, and with Jimmy Carter, who used it in 1978 to safeguard 56 million acres of spectacular federal land in Alaska after Congress balked at enacting protective legislation.

The surprising fact is that FDR used the Act more often than any president, when measured by the number of monument proclamations, and (except for Carter and Clinton), when measured by the number of acres protected. FDR issued 29 national monument proclamations, compared to 19 for Clinton and 14 for TR.

Beginning in 1999, when we got deeply in the business of preparing new national monument recommendations to President Clinton, Secretary Babbitt took to carrying in his pocket a 3 × 5 index card on which he kept a running tally of the acreage set aside under the Antiquities Act by TR, by FDR, and by Clinton. At opportune moments in encounters with President Clinton, Babbitt would surreptitiously hand him the card, to remind him how he was doing compared to his predecessors. Now I am not so cynical as to suggest that President Clinton was motivated to protect some worthy areas of federal land simply by competition with his illustrious predecessors, but we are all human and, as Roger Kennedy said earlier, it is never too late for us to do better.

With that as a springboard, I want to focus more broadly on FDR's impact on federal lands—that one-third of all the real estate in America, which is owned and managed by the national government. Much of this national patrimony is managed through great national systems—the national parks, monuments, forests, wildlife refuges, and other conservation areas. These systems are models now being widely copied around the globe. But few know just how instrumental FDR was in fundamentally shaping them.

Consider the so-called crown jewels: When FDR took office, there were some scattered national parks, mostly in the scenic mountains of the West, but there was no truly national system. In 1934 FDR issued an executive order that fused several disparate collections of areas into a genuine system. These included national monuments formerly managed by the U.S. Forest Service, national battlefields managed by the military, and national historic sites. In so doing he broadened the focus of the National Park Service to include cultural as well as nature preservation.

FDR also strongly supported expanding the system to include different faces of nature. Early in his tenure, the Florida Everglades (distinctly not mountainous—its highest point being only 8 feet above sea level), was brought into the system. Preserved for its biological richness, this 1.4 million-acre landscape-scale park blazed a path that has been followed many times since.

Other large new parks designated on his watch, such as Big Bend in Texas, added further diversity. FDR became deeply personally involved in a significant expansion of the Olympic National Park in Washington, extending it beyond its original "rocks and ice" character to include a magnificent rainforest. He also helped bring Cape Hatteras National Seashore into the system, the first of what has today become a large number of national seashores. Several national monuments were expanded and converted into national parks during his administration, including Zion, Death Valley, Joshua Tree, and Capitol Reef.

FDR's land legacy goes well beyond the national parks. When he took office, almost all the national forests were in the West, despite the fact that

Congress had, in the Weeks Act of 1911, authorized the acquisition of cutover forest lands in the east for restoration and inclusion in the system. FDR personally pushed for a major expansion of this program, which resulted in the acquisition of some 20 million acres of forest land in the East, Midwest, and South. The record shows that FDR—who, after all, considered himself a forester—was purposeful in this expansion. He wanted to make the national forest system truly national, to demonstrate governmental leadership in forest restoration across the country. FDR's additions now provide important biological diversity and recreational escape for many population centers in the eastern half of the country.

Then there were the national wildlife refuges. While Teddy Roosevelt came up with the idea three decades earlier, here too FDR led a vast expansion, overseeing the creation of 150 new refuges (against 100 in existence when he took office), and more than doubling the acreage in the system. The Duck Stamp Act, which continues to this day to provide funds for acquisition of new refuge lands, was also enacted on his watch.

It was, however, with respect to the so-called public domain lands that FDR may have had the most profound, if most obscure, impact. These were the leftover lands, the largest chunk of federal real estate—more than two hundred million acres—and the least known. The nineteenth-century policy of divesting these lands out of national ownership was still in place when FDR took office. Thirty-six million acres of these lands had been privatized under the homestead and related acts in the 1920s alone. In fact, not three years before he took office, his predecessor had offered the states title to all the remaining federal public domain (minus the minerals) as an outright gift. While Herbert Hoover was a constructive conservationist in some ways, he wanted no part of this vast real estate.

Here FDR gets all the credit for preserving these lands, and the natural riches they contain, in federal ownership. He began by throwing his support behind efforts to reform federal policy toward livestock grazing on these lands, then their only significant use. Shortly after he signed the Taylor Grazing Act into law in 1934, FDR boldly "withdrew" practically all the public lands from further disposition by executive order. This effectively ended large-scale privatization and disposition. By this action, FDR secured in governmental hands a number of outstanding areas, paving the way for them to become national parks, wilderness, and other protected areas, prized for their environmental, scenic, cultural, and recreational values. In fact, many of President Clinton's national monuments were proclaimed on lands preserved in public ownership by FDR.

A final observation: It is true that, like much of the rest of FDR's conservation legacy, these efforts were mostly completed by 1939. But FDR did not lose interest in such matters with the outbreak of World War II. He played

a key role in bringing Kings Canyon in California into the national park
system in 1940 with a special, unprecedented mandate, to remain a wilderness, preserved in its wild state. Then there was the Jackson Hole National
Monument. In 1943 Laurence Rockefeller, tired of waiting for Congress to
agree to protect lands he had purchased along the front of the Grand Tetons
and offered to donate to the United States, signaled he would sell them off.
FDR rose to the occasion, using the Antiquities Act to accept title and proclaim these lands (along with 173,000 acres of national forest lands) a
national monument.

The Wyoming congressional delegation protested loudly (one senator calling it a "foul Pearl Harbor blow" just 18 months after December 7, 1941)
and pushed through a bill to repeal it. FDR vetoed it. A young local county
commissioner named Cliff Hansen and actor Wallace Beery drove 600 livestock across the protected lands in protest. Years later Hansen, who went on
to serve several terms as U.S. senator, acknowledged that his opposition had
been a huge mistake, and that FDR's vision had saved the magnificent
Jackson Hole area from the ruin of overdevelopment.

Throughout the war, FDR resisted opening up protected federal land
areas to exploitation, rejecting such proposals as harvesting spruce in
Olympic National Park for aircraft construction. His wartime defense of conservation stands in sharp contrast to our current president, who holds out the
false hope that we can drill our way to energy security by opening protected
areas of federal lands.

In short, FDR's conservation legacy is not just in ideas and programs. One
of his greatest endowments to succeeding generations of Americans is what
he did literally on the ground—his profound impact on national landholdings and management.

CHAPTER 8

REFERENDUM ON PLANNING: IMAGING RIVER CONSERVATION IN THE 1938 TVA HEARINGS

BRIAN BLACK

HISTORIANS HAVE LONG NOTED THE REVOLUTION BROUGHT BY THE NEW DEAL to the realms of politics, social planning, and even the administration of natural resources. More complicated, however, is deducing the degree to which these ideas can be attributed to the president responsible for implementing and administering them. A close examination of the Tennessee Valley Authority (TVA), and of 1930s river development in general, reveals that Franklin D. Roosevelt's (FDR's) thinking did influence this new effort to rationalize and manage natural resources. In fact, the TVA possibly reflects FDR's early environmental thinking better than any other New Deal agency. This is not to say that the TVA, like many other New Deal agencies, did not change with time. During World War II, for example, the need for power generation usurped many of the founding priorities of the river development agency. Today, the TVA is recognized as one of the greatest environmental exploiters in the nation—which, of course, is a legacy we likely do not wish to claim for FDR. But if we focus our view more on the TVA's early years, specifically between 1933 and 1938, we find that it connects more directly to a legacy of environmental design and policy making—a legacy that is clearly tied to the early environmental thinking of FDR.

Let us begin by considering the priorities of the TVA between 1933 and 1938 in a chronological vacuum, for it is in the roots of a new era in

American land management. In reference to the agency's team of three directors, Garet Garrett explained to *Saturday Evening Post* readers in 1938: "Never before had the Congress delegated so much power to three men. . . . It was the power to set up a regional government within the United States; one that nobody elected, one that could not be changed or voted away by the people on who [m] it acted."[1] The TVA Act stipulated that the organization would construct and operate a system of multipurpose dams and reservoirs on the Tennessee River and its tributaries to promote navigation, control flood waters, and provide facilities for the generation of electric energy. However, the Act did not place limitations on the methods the TVA used to arrive at these ends.[2] The agency and its directors enjoyed free reign through the mid-1930s but ran directly into public questioning in 1936. These questions would evolve into a full-blown governmental inquiry, including 1938 congressional hearings and a ruling by the Supreme Court on the agency's right to exist.

The public's curiosity and uncertainty over this revolutionary new fashion of using nature plays out in the congressional hearings that were held in 1938 to decide if the TVA was successful and even legal. The TVA, of course, stood as the great national experiment in regional planning and river development; therefore, these ideas and concepts stood on trial with the agency. The 1938 hearings clearly serve as a referendum on the planning ideas behind the TVA—ideas that clearly emanated from the oval office.

PLANNING FUNCTIONAL NATURE

In an interview a few years ago, the late pioneer landscape designer Ian McHarg commented, "When it was established in 1933, [the] TVA represented the most utopian ideas of resource management found in the nation." From the moment of the TVA's conception, FDR realized that an effort to redesign a river valley was monumental in scale, and that if it was successful, it would provide a model that would reinvent ideas of land planning. More important than kilowatts and fertilizer preparation was the *idea* of the project, for the efforts to establish the TVA brought a model of conservation planning to the American landscape unparalleled in its scope. More than any other agency, though, TVA marketed an extremely technologically based model of conservation to the American public.

The TVA Act of 1933 called for the U.S. government to finance, plan, and carry out the revitalization of a depleted region by constructing a series of dams along the Tennessee River that would harness the river's potential for generating power while also tempering its flow to prevent flooding. The first TVA project, Norris Dam, cost $34 million to erect. During the 1930s, the United States invested $300 million in TVA projects, creating eight dams along the Tennessee.[3] Taming the river led to many ancillary duties such as relocation, town planning, and recreational administration. With these additional duties in mind, the TVA conceived of its purpose in a much more far-sighted fashion than simply constructing dams. Once the river was controlled and soil runoff eased, TVA planners set out to make sure that the land would not be depleted again. This was to be accomplished through education, both regional and national. One of the most instructive examples of conservation was the landscape created by the legions working under the agency's three directors, Arthur E. Morgan, David Lilienthal, and Harcourt Morgan.

While scientific planning need not involve aesthetics, the TVA's designs stressed visual appeal to symbolize a new ideal landscape and, in later years, to stimulate recreation.[4] While the New Deal introduced the public to these new ideas of ecology and conservation, it was still bound to conform to public support. The way of doing so, strategists decided, was to give Americans the reasons behind the federal actions—to educate them in conservation. Whereas other policy initiatives called for publicizing certain causes or issues, conservation required that Americans be given an introduction to the natural sciences. In the aggressive way that FDR would inform the public of each New Deal effort, conservation was made public through print, film, and photograph.[5] While the TVA's construction efforts were confined to the watershed of the Tennessee River, the publicity generated by the agency was directed at the entire nation.[6] For many Americans, the TVA was the first symbol of conservation in practice.

Once the TVA began purchasing land and directing residents to alter land-use practices, however, public support for the project began to wane, particularly among the local inhabitants. This ambivalence could also be found in the press. In an article written for Collier's in 1933, for example, Walter Davenport noted: "There are those who know it can't be done, because it has never been done [before]." But, he continues, "Roosevelt's plan is . . . a beautifully natural plan, one that must develop gradually at great initial expense." FDR's is the "dream with teeth in it."[7]

How much of TVA's conception can be attributed to FDR? At the very least, FDR must be credited with the personal selection of Arthur E. Morgan. Morgan's fiery personality in many ways brought on the congressional investigation of the late 1930s; however, his combination of utopianism and river engineering defined the TVA's early years. Morgan raised the TVA from a hydroelectric project to that of a social experiment exploring humans' relationship to the natural environment. The rise and fall of Morgan's influence in the 1930s appears not to have injured FDR's support of the agency— though it has been carefully noted that he possessed unique admiration for Morgan and his ideas. Clearly, though, the departure of Morgan in 1938 signals a wholesale shift in emphasis for the TVA from emphasizing land planning to power generation and sales.

FDR heard of Morgan through James M. Cox, the former governor of Ohio, who ran for president with FDR as his running mate in 1920. During Cox's tenure as governor, Morgan had developed the innovative Miami Conservancy river development project. He had taken over the presidency of Antioch College, with an emphasis on developing its engineering school. Cox would later recall: "I remember well remarking to the President that Morgan would be honest and efficient but had no patience with politics in matters such as this." Historian Roy Talbert writes: "That is part of the enigma that is Roosevelt. Part of him liked the Utopian. Within twenty-four hours after discussing Morgan with Cox, Roosevelt offered the Antioch College president the position."[8] When Roosevelt invited Morgan to Washington in April 1933, the engineer assumed it was to discuss his ideas about expanding Miami Conservancy. Instead, Roosevelt abruptly offered him the controlling directorship of the TVA. In agreeing with Morgan's lone demand, FDR reportedly slammed his hand on his desk and emphatically promised that there would be no politics in the project.

After their meeting, Morgan soon realized that the TVA Act was still in congressional conference committees, so he immediately began efforts to have a say in its drafting. Most important, Morgan wished to make certain that the TVA would not rely on Army Corps of Engineers for construction. Such a cooperative endeavor would considerably limit the connections of cohesive watershed planning that he wished to implement. When it was complete, the TVA Act prioritized multipurpose river basin planning in a revolutionary vision that teamed engineering with ecological understanding and social design.

A brainchild of Senator George C. Norris, the TVA also grew in the mind of FDR immediately when he took office. Infusing the energy production idea with the conservation ethic that he had fostered throughout his life and early political career, FDR required an expert engineer as well as a utopian visionary to lead the TVA. Morgan may well have been the only individual in the world with this combination of river engineering experience and interest in social planning.[9] The scale of his aspirations required an entire river valley; however, even this at times appeared insufficient. A visionary designer, Morgan sought to create a landscape and a human culture that worked as one—he sought the middle ground that historians of technology have contemplated from Leo Marx to Lewis Mumford. The TVA would become a defining experiment in the idea of regional planning, a forerunner of environmental planning.

The regional planning movement grew out of the modern field of human engineering that Norris and others believed could resurrect a downtrodden portion of society.[10] For his Chief Planner, Morgan, the utopian dam builder, resisted pressure by Lewis Mumford and others to hire a regional planning enthusiast; instead, Morgan chose Earle Draper, a landscape architect. Unlike much of the New Deal's landscape planning, the TVA, under Draper's guidance, would not abandon aesthetic concerns. Draper raised the discipline of landscape architecture to its apex of development and application, and made the TVA the culmination of the field's "early initiative in planning."[11] While director of the TVA's Division of Land Planning and Housing, Draper supervised architects, engineers, social scientists, planners, geographers, and a core of nearly 20 landscape architects. The actual design of the physical landscape became a crucial portion of planning—at least partly because of its symbolic potential.

Morgan and Draper's TVA represented a national trend. As journals for planning were founded and 42 State Planning Boards were established under the guidance of one National Planning Board in the 1920s, planners and landscape architects found themselves in great demand.[12] Landscape architect Phoebe Cutler writes that despite such a national trend, once the TVA was established in 1933, it became the "spiritual home" for all of landscape architecture.[13] The magical hand of federal authority gave designers carte blanche to make this valley an ideal, no matter if the plan necessitated moving families, towns, or forests to more convenient locations. Unlike many previous products of landscape architecture, however, this landscape was intended as a multiuse area based in conservation.

A PROBLEM OF VISION

Before and during these internal conflicts, the TVA fielded many legal charges against its right to exist, and the limits of its jurisdiction. The legal fees ran into millions of dollars, but also contributed to public ridicule of the TVA's political viability. The dissolution of the TVA's public image can be seen in a 1938 editorial from *The New York Times*. The editorial is set in the courtroom of the Mad Hatter, of *Alice in Wonderland* fame. The Hatter immediately rebukes Alice for naïve attitudes: "That's not the way the world is run any more. . . . We don't do things today because they are essential or because our heart is really set on them, but only because they are incidental to something else." Alice has no idea what the Hatter is talking about. The Hatter explains by using a notable example:

> It is quite simple, Alice. TVA is going to spend $500,000,000 to develop a lot of waterpower in the Tennessee Valley. But do we really want the waterpower? Oh, no! The waterpower is only incidental to improving navigation on the Tennessee River, which can be done for about $75,000,000. It is also incidental

to flood control, which can be done for about $50,000,000. So you see, my dear, that the modern way is to spend an awful lot of money incidentally to [do] something else that could be done for one-fifth the cost.[14]

The Hatter concludes by comparing this example to a college girl's itemized expense list: "Room rent, $25. Clothes, $10. Tips, $2 . . . . Trolley fare, 35 cents . . . . Incidentals, $276.50."

At times, the TVA seemed to make it painfully easy for the public to take up TVA-bashing. Such a furor of interest was generated over the infamous jackass episode. In March 1938, Senators McKellar of Tennessee and Senator Bridges of New Hampshire found themselves at odds about the TVA's policies. Senator Bridges, in his rage, decided to tell the Senate a story about a jackass—not just any jackass, but the "perfect jackass." He made the claim that the TVA had spent $4,500 to search for and buy a "perfect jackass" that would serve as sire of a new race of "super jackasses." " 'Came the spring, the season of daffodils and young love,' Mr. Bridges continued. 'Romance blossomed on the broad meadows of the Tennessee Valley. But the TVA's jackass just wasn't interested. The experts gathered, decided to sell the animal. They got $300 for him.' "[15] A few days later newspapers around the country extended the ridiculous tale by reporting that the TVA's jackass was dead. A farmer reported he had purchased the animal from the TVA for $350 and was offered $800 by another farmer shortly before the animal died.[16] The TVA waited a week before commenting on the whole episode. When it did, *The New York Times* reports, "The [TVA] pushed aside the intricate problems of one of the country's greatest engineering projects long enough today to unravel the mystery of a jackass." The story explained that the cost had been confused with the cost of four mechanical jacks. The TVA claimed to have paid only $290 for the animal, and sold it at an $80 profit.[17]

To the American public, the TVA may have seemed to be in control of its valley, but throughout the late 1930s its role in national politics grew more and more embattled. Primarily, divisions had surfaced among the three directors. Originally, the unique, three-way division of leadership power was intended to keep individual desires in check. In the Board Minutes from August 22, 1933 the responsibilities were laid out for Harcourt Morgan to oversee agriculture and fertilizer efforts, Lilienthal to oversee the distribution of power, and A.E. Morgan to, among other activities, integrate various parts of the work into a unified program.[18] As the Chair and founding director, A.E. Morgan had imposed his personal vision on the agency and clearly let it be known internally that one of the other nominees, David E. Lilienthal, did not cohere with his viewpoint. The matter became public knowledge during late 1935 and would remain a well-known portion of the TVA through Morgan's departure in 1938.

At times, it seemed that the work of the TVA was directly intertwined with its leadership difficulties. Willson Whitman describes the situation in a 1938 *Harper's* article:

> Three honest men have been disagreeing in the Tennessee Valley. When the Tennessee Valley Authority was first established it was a question among people who knew the country whether any three men could lick the job before the Tennessee [River] could lick them. The first round has now gone to [the] Tennessee [River].[19]

The TVA's effectiveness, Morgan would argue, required the universal commitment of the three directors to the overarching goals and responsibilities of the agency; the directors must not merely oversee the portions of the undertaking that were stipulated in the original act. From the outset, Morgan believed that Lilienthal had tunnel vision for the legal and power maneuvering necessary for the agency to dominate regional energy development.[20] The more utopian portions of the experiment, such as the construction of the town Norris, Tennessee or the overall balance of the river valley's ecological integrity, grew from Morgan's initial plans for the TVA. Such plans often grated against the priorities of the power division and also made it difficult for the public to fully comprehend the work that the TVA pursued. These internal difficulties were leaked to the American public in 1936, when David Lilienthal's term expired. A.E. Morgan made unofficial overtures to FDR and Senator George W. Norris to request that Lilienthal's term not be renewed. To Norris, Morgan stressed that the issue of power generation was not his major concern; instead, he was most disturbed by his "lack of personal confidence" in Lilienthal.[21] It can be inferred that he lacked confidence that Lilienthal would fairly consider the full breadth of the TVA's utopian roots.

In his personal recollections, Morgan specifically states that FDR assured him that Lilienthal would not be renewed. Clearly, though, the national press had openly discussed the disagreement as well as the personalities of each director. In Morgan's estimation, Lilienthal had generated much of the publicity in an effort to back his way into reappointment. Regardless, FDR's personal letter to Morgan explaining the renewal of Lilienthal is clear in its inferences. FDR urged him

> . . . to consider not only the big fundamentals involved, but also two other considerations: First, that the TVA is the first of, I hope, many similar organizations that we are all in the formative period of developing administrative methods for a new type of government agency; second, my own personal problems. In regard to the latter, I have told you of the somewhat heavy load which is on my shoulders at the present time. I ask your sympathetic consideration.[22]

Once the disagreement was in public view it might influence public feelings of the New Deal, FDR, and, of course, the TVA. These are the larger issues to which the savvy FDR alludes in his correspondence.

Lilienthal's renewal brought with it calls for Morgan's resignation. In rejecting such an idea, Morgan demonstrated the significance of the stakes at hand. Control of the TVA was much more than a trophy of power and prestige; indeed, he had come to believe that the very idea of balanced river development was held in sway by the agency's future. The TVA's status as a national experiment that could lead to other regional authorities only made his stand more imperative. In a telegram from January 1937, Morgan urges FDR: "In the development of a federal power policy it is important that the relation of such a policy to problems of unified river control be not over-looked. . . . Unified river control demands that river systems be operated as units for all associated purposes including navigation, flood control, and power."[23] FDR's personal response passed below the radar of political consideration to consider the vision of environmental planning that the two men had shared in the earliest days of the TVA: "You are right about the coordination of the problem of a river as a whole. I hope you will run in to see me very soon."[24] Morgan, however, had lost faith in the president's ability to commit to the integrity of their original plans.

In a fateful act, Morgan decided to display his views of river planning to a public audience. His article "Public Ownership of Power" would be released later in 1937 to readers of the *Atlantic Monthly*. Regardless of his disclaimer near the article's close explaining that the article is based on his own opinions and that he is only a "minority member of the Board of Directors of TVA," Morgan had severely compromised the agency and the strict ethics enumerating the directors' abilities to speak publicly on such subjects.[25] After a lengthy, technical discussion of the evolution of electric utilities and the role for the federal government in managing the supplies of such basic resources, Morgan writes:

> One of the largest public projects for supplying electric power is the TVA. The TVA is an organization for varied purposes, largely centered around the unified control of the Tennessee River system for navigation and flood control, as well as for power. The flood control and navigation programmes of the TVA are not "a masquerade . . . ." If the TVA act is fairly interpreted and administered, it can mark a great advance in the planned and orderly development of a great river system.[26]

Following the article's release, Harcourt Morgan joined Lilienthal in openly questioning A.E. Morgan's competency to act fairly on behalf of the TVA. Their official memorandum to FDR carefully enumerated A.E. Morgan's inability to cooperate with the other members of the organization.

A.E. Morgan quickly sent charges of dishonesty, malfeasance, and lack of integrity back at the other two directors. His refusal to resign left the president with little choice but to allow a congressional investigation to play out. Such a public display, of course, would not limit itself to the actual problems at hand; instead, the agency would need to explain its efforts and organization to the American public.[27]

Leading the criticism of the TVA (and by association of regional planning) since 1933 had been southern utility companies, particularly coal suppliers. At least two of the legal cases filed by these companies rose to be heard by the U.S. Supreme Court. The second case came to the court just as the congressional hearing began over the TVA. The congressional committee met repeatedly throughout 1938, including a five-day tour of the Valley by the entire proceeding. Each turn of the hearing, of course, tapped into the public's curiosity about the inner workings of the TVA. One hundred and one witnesses were heard, generating 15.5 thousand pages of testimony and 488 exhibits.

In his testimony, A.E. Morgan surprised many onlookers by recommending that the leadership structure of the TVA should be reconsidered. Presently, he stated, the TVA is "too varied and far-flung to be well administered by a single administrator or a single general manager under a board." The TVA Act had specifically directed "the subordination of power to flood control and navigation." The actual occurrences of construction and development had reconfigured the TVA by "accident." To solve such problems, Morgan suggested that directors must be influenced by the nature of the activities of the TVA as defined in the TVA Act.

> In some respects this division of activities is a matter of practical convenience. However, in respect to separating unified river control from the detailed distribution, marketing, and sale of electric power, the matter is *not* just one of convenience, but of vital public safety. Unified river control for all useful purposes is a great advance in engineering and in public safety. The Tennessee River project can mark the way for unified treatment of other river systems. It would be a matter of regret to see that undertaking abandoned or deformed. Yet its success depends on the utmost integrity of administration, and on the fullest assurance that flood control storage space shall be inviolate for flood control and shall not be infringed upon for immediate power profits.[28]

Morgan had little doubt that the very ideas behind the TVA are what the hearings actually wished to air out. With this in mind, the other directors performed similarly.

With great pride Harcourt Morgan repeatedly referred to the TVA as "This great experiment." He referred to the TVA as a "Fundamental program" and then went on to explain that it deals with fundamental causes at their roots. The TVA program, he explained:

> . . . is further fundamental with respect to the very subject matter involved. That subject matter is the basic facts of nature upon which the entire structure of our civilization depends. I mean the water, the land, and the energy in that water, and the fertility in that land. These are the things that nature has offered to man and that constitute the foundation upon which man fashions his institutions, social, economic, political. . . . Those of us who believe in democracy have confidence that the people will find and fashion those institutions best fitted to their needs through their own initiative—provided those gifts of nature which I have mentioned are conserved, and wisely utilized. . . . The water and the land together are the bases of all biologic life—plant, animal, and human. Land without water is uninhabited desert. Nor will water without land support human life. Where both are plentiful together, there has always been the seat of abundant life and the growth of civilization. . . . These facts of nature—land and water—are not separable. The unwise use of the land by the destruction of its crop and forest cover hastens destruction of that soil by the water falling from above . . . These facts give rise to the need for basic adjustments. It is with these fundamentals of nature that TVA is concerned— whether they shall become cumulatively the enemy of human life, or whether they shall be conserved, their continuing benefits captured, and the water controlled so that the people may build their lives upon them.[29]

The ethic of conservation seemed to bring the Morgans together even in their moment of greatest division. Most surprising, Lilienthal spoke in similarly grand terms. In reference to the TVA, he said: ". . . though it is an agent in a new form, it is engaged in problems which are as old as the Republic, the problem of conservation, the efforts to conserve our natural resources of soil, of forests, or water and minerals, is one that is rooted deep in the needs of American life."[30]

Regardless of the committee hearing from over 1,000 others, the ideas and principles of these three witnesses were the major exhibits in the TVA hearing. The other witnesses included technical members of the TVA discussing day-to-day operation and representatives of coal companies claiming to be adversely effected. The ten-member committee summarized its findings as follows: "The value of the Authority lies in providing the only method by which all the uses of the river can be brought within reach of the public."[31] Again, the refrain of the hearing was unified river development. And yet when FDR relieved Morgan of his duties at the hearings' close, the TVA had lost its greatest instrument for bringing together the disparate portions of its mission.

While the comments of the three directors returned to the idealism in which Morgan and Roosevelt had established the TVA, Supreme Court justices were simultaneously weighing the legality of government involvement in such efforts.[32] Clearly, the hearing, the dams, and all that had been accomplished in five years in the valley were moot if the TVA's very reason for being was deemed a conflict of interest. Nineteen power companies joined in the suit filed in 1936 to allege that the TVA, under the false guise of navigation, flood control, and national defense, was entering into a vast program of power production. Decided in a three-judge district court, the decision favored the TVA on every factual point.[33] The Supreme Court's hearing of the case actually only grew from the effort to establish standing to sue. In essence then, the decisions sustained the good faith of the TVA to uphold the mandates of the 1933 TVA Act. The Supreme Court dismissed the case and in late 1938, the TVA was—for the first time since its establishment—legal and free to operate. However, it was not the same agency that had come into the 1938 referendum on planning.

THE IMPLICATIONS FOR ENVIRONMENTAL PLANNING

New Dealers went into the 1938 hearings with grand hopes of parlaying the TVA's work into at least twelve other river authorities to be operated throughout the nation. Orchestrating the development of regional planning entities, the Natural Resource Board was a vague agency that changed names at least twice. While the ideas remained valid, the unified effort and authority seemed impossible to maintain. Attempts to do so during the early 1940s ended negatively. While the fields of river engineering and landscape architecture and planning would be forever changed by the early efforts of the TVA, applying such ideas to an entire river basin in the TVA fashion would be lost. Harcourt Morgan had been appointed temporary chairman of the TVA directors, and he acquiesced this position to Lilienthal following FDR's dismissal of A.E. Morgan on March 22, 1938. The issues of national defense, of course, rapidly usurped any other agenda for the TVA. However, Lilienthal would openly allow the agency to emphasize power production in a fashion that the elder Morgan found distasteful.

In the final estimate, it appears that A.E. Morgan—for all his faults of ego, and so on—was that unique engineer with the ability to connect his technical projects with a grander concept. The concept of ecologically friendly design endured, but the TVA mold would not be followed again in the United States[34] The American public maintained its infatuation with the TVA even after the referendum. The 1930s ended with the TVA making every effort to move forward. The cover story in a 1939 *New York Times Magazine* lifts the curtain on the TVA's self-proclaimed "second phase." With the feuds and legal wrangling concluded, the TVA's legal status cleared by a 1938 Supreme Court ruling, and the projects along the river completed, the TVA began its administrative phase.

Beyond the TVA, however, ideas of regional planning had dramatically changed. As the TVA moved forward, the ideas of its founding—particularly that of unified river planning—graduated from the inhibitions of politics to incubate within the scientific and design practitioners. For instance, in 1939 *Harper's* published "Science and the New Landscape" by Paul B. Sears, the revolutionary ecologist who wrote *Deserts on the March* in 1935. He explains to readers that landscapes are no longer inert or changeless objects. The sciences of ecology and anthropology, which made up the foundation of TVA practice, "have taught that man himself is not a watcher, but like other living things, is a part of the landscape in which he abides. This landscape, including its living constituents, is an integrated whole."[35] He concludes by saying that scientists of the future must continue to work with an awareness of social and economic processes—but there is another hurdle: "The scientist must be aware of the relation of his task to the field of aesthetics."[36] By this point at the end of the 1930s, the TVA, its dams, and the re-made river valley were a symbol of this marriage that Sears observed—the middle landscape. A new aesthetic of efficiency had taken shape with the modern landscape serving as its product and symbol.

As the river and watershed was manipulated, the Tennessee Valley project became an early example of the environmental planning that would reemerge in the 1960s. If one merely looks at the technological landscape that is created, she may miss the intent behind it. During these early years, the environmental ethic of Morgan and FDR remade the Tennessee Valley as an instrument for social, economic, and ecological change. Actively managing the landscape through technology of the modern era, planners created a symbolic landscape of conservation in action. Historians must be able to look beyond a landscape's instrumentalization to reconstruct its rationale. In the case of the TVA, we learn that the technical landscape could have an environmental plan at its core. This plan could contribute mightily to altering the nation's environmental ethic.

NOTES

1. Garet Garret, *Saturday Evening Post,* May 7, 1938, Ibid., p. 9.
2. The TVA Act, 1933, as amended.
3. Figures quoted from *The New York Times Magazine,* August 20, 1939.
4. A *Collier's* article in 1935 explains that Americans would like to be "as careless of the future [resources] as our forefathers were, but we can't do it. The reasons are apparent."

The landscape designers of TVA were commissioned to conceal and integrate the hardware of resource conservation. Cutler calls the planned valley the "ultimate watershed demonstration area." The New Deal presents the high point of this effort to rationalize natural resources in the United States during the modern era. In an effort to better understand this complicated era of change, my argument today urges environmental history to bring the historical study of modernism and the technological landscape more into its purview. Traditionally, the landscapes of modernity have not interested environmental historians; worse, many of us have viewed such managed landscapes with disdain reserved for shopping malls and parking lots. But environmental historians must reconcile the role of technology and planning so that the interwar era can assume its rightful position: situating and, in many ways, defining what we call the modern environmental movement of the twentieth century. Most important, these years situate the federal government as a significant device for environmental mediation and management.

5. Investigating the New Deal's interest in validating itself through the popular culture is not the specific purpose of this analysis. However, the consideration of this agenda is based on a variety of dynamics blending together during the 1930s. Warren Susman writes, "The photograph, the radio, the moving picture—these were not new, but the sophisticated uses to which they were put created a special community of all Americans [in the 1930s] unthinkable previously." *Culture as History* (New York: Pantheon, 1984), 160. Additionally, William Stott writes simply, "[FDR] had a documentary imagination: he grasped a social reality best through the details of a particular case." *Documentary Expression and Thirties America* (Chicago: University of Chicago Press, 1986), 94. These are two of the most important cultural dynamics that created an era of politics that openly used many forms of propaganda and popular culture to take policy explanations directly to the public, often bypassing traditional political avenues.

6. This terminology was a standard reference to TVA during the 1930s, suggesting a connection of the United States, also considered a "grand experiment." FDR often referred to TVA efforts in this way.

7. Walter Davenport, "The Promised Land," *Collier's,* June 3, 1933, pp. 13 and 37.

8. Talbert, 84.

9. See*** *FDR's Utopian* for an excellent discussion of Morgan's writing and thinking on utopian literature.

10. Many sources offer this insight. The most interesting might be Thomas Hughes's (New York: Penguin, 1990) *American Genesis,* which places TVA as a monumental application of American technological "enthusiasm" between 1870 and 1970. Hughes argues that TVA provides a model of planning and bureaucratization that permeates the American corporation, government, and even everyday life.

11. Phoebe Cutler, *The Landscape of the New Deal* (New Haven: Yale University Press, 1985): 135.

12. These journals include *The Planner's Journal* and *Planning and Civic Comment.* Draper was the first editor of *Planners Journal.*

13. Cutler, *The Landscape of the New Deal,* 135.

14. *The New York Times,* January 30, 1938, IV, 8:4.

15. Ibid., March 10, 1938, 4:3.

16. Ibid., March 14, 1938, 12:7.

17. Ibid., March 24, 1938, 3:2.

18. TVA Board Minutes, August 22, 1933.

19. Willson Whitman, "Morgan and Morgan and Lilienthal: The Personalities in the TVA Wrangle," *Harper's* (September 1938): 352.

20. This is particularly evident in the materials that Morgan prepared for the 1938 hearings in which he retraced his interactions with each of the parties involved.

21. Letter of August 13, 1936.

22. Personal correspondence, FDR to A.E. Morgan, May 15, 1936.

23. Telegram, AEM to FDR, January 27, 1937.

24. Personal correspondence, FDR to AEM, January 29, 1937.

25. These ethical guidelines for directors are collected in the Minutes of the Board Meetings, 1933.

26. *Atlantic Monthly* (September 1937).

27. Readers may also wish to consult North Callahan, *TVA: Bridge Over Troubled Waters* (Stanford CT: A.S. Barnes & Company, 1980), which contains an outstanding retelling of this episode accompanied by personal interviews with A.E. Morgan.

28. "A Summary of 'The Power Operations of the Tennessee Valley Authority'" December 1938, submitted to Congress 13.

29. Hearing testimony 6–9.

30. Ibid., 912–13.

31. Ibid., 234.

32. For an extended treatment of these legal cases, see *TVA: The First Twenty Years,* edited by Roscoe C. Martin, 29–34.

33. *Tenn. Electric Power Co. v. TVA,* 21 F. Supplement 947 (Electricity Department Tennessee, 1938).

34. It should be pointed out that the mold of TVA would be exported throughout the world.

35. Paul B. Sears, "Science and the New Landscape," *Harper's* (July 1939): 207.

36. Ibid.: 216.

CHAPTER 9

"FDR AND ENVIRONMENTAL LEADERSHIP"

JAMES R. LYONS

INTRODUCTION

JAMES MACGREGOR BURNS, IN HIS PULITZER PRIZE WINNING TREATISE, *LEADERSHIP*, begins his prologue with the following statement, "One of the most universal cravings of our time is a hunger for compelling and creative leadership."[1] If this is true today, or at the time that Burns completed his seminal work on leadership in 1978, it was most certainly true in 1933, when economic crisis and natural resource destruction combined to cast a pall over the American landscape. Leadership was desperately needed. Franklin D. Roosevelt (FDR), in so many ways, has been heralded for the leadership he and his administration provided to a nation in despair. But few have adequately recognized the leadership provided by FDR in restoring the nation's natural resources, in protecting special places, and preserving the environment.

In thought, word, and deed, FDR and the cadre of environmental leaders who served him in office, demonstrated a commitment to environmental protection that may be unparalleled in American presidencies. While the accomplishments of his distant cousin, the twenty-fifth president of the United States, Teddy Roosevelt (TR), are often heralded as among the most significant of the American presidents, FDR's accomplishments are as substantial and, in some respects, even more broad and profound than TR's.

There are a number of reasons for this. First, FDR's conservation accomplishments affected both public and private lands. In so doing, FDR's conservation efforts promoted both the preservation of unique and important landscapes in the public domain and the stewardship of lands in private ownership. In a sense, this was the first effort to promote what we now term "landscape level" conservation—an essential foundation for the preservation of biodiversity, the protection of watersheds, and the sustainable use of natural resources.

Second, FDR brought new, creative and visionary leadership into the federal government, and, in particular, into the departments and agencies with primary responsibility for protection of the American environment. Through the leadership of Harold Ickes, in the Department of the Interior; Henry Wallace, secretary of the Department of Agriculture; Hugh Hammond Bennett, the first chief of the Soil Conservation Service, and Jay N. "Ding" Darling as director of the Bureau of Biological Survey, FDR demonstrated a capacity to identify and encourage effective conservation leaders.

Third, FDR demonstrated a capacity to promote the institutional changes needed to construct new approaches to conserving the nation's land and water resources. He did so both in the establishment of new agencies, such as the Soil Conservation Service in the Department of Agriculture, and in advocating new programs and policies to promote improved conservation. The Agricultural Adjustment Act of 1933, one of FDR's first initiatives, introduced the novel concept of paying farmers to not cultivate land but, instead, to dedicate such lands to conservation uses.

Fourth, FDR demonstrated that a healthy, productive environment and a sound economy could go hand in hand. The Civilian Conservation Corps (CCC), one of FDR's most notable conservation achievements, was established for the dual purpose of providing employment for millions of unemployed Americans and the restoration and protection of the nation's natural resources.

Finally, as illustrated by the aggressive and effective conservation agenda of the Roosevelt administration, FDR demonstrated that environmental protection need not be sacrificed during a time of national crisis. To the contrary, FDR's commitment to conservation remained steadfast despite concern for a failing economy and the impending conflagration in Europe, Asia, and the Pacific. In fact, this presents an interesting contrast to the situation we find ourselves in today.

As extended warfare in Iraq and Afghanistan intensified, and talk of a perpetual "War on Terror" persists, the concern of the Bush administration for the environment languishes. Critics have suggested that the focus of the Bush administration on warfare is intended to distract the public from domestic ills, including environmental degradation. Others have pondered if concern for "national security" hasn't, in fact, been used as a rationale for plundering resources and excusing the negative environmental consequences of environmental policies advocated by the administration. As one example, one of the first products of the Bush administration was the report of the National Energy Policy Development Group recommending a National Energy Policy to the president. The report emphasized the need to promote additional domestic energy supplies as a means of promoting "U.S. energy security" and served as the basis for the president's energy security legislation.[2] Critics of the plan have pointed to the legislation and other administrative initiatives on the part of the Bush administration as evidence that the environment is being sacrificed in the name of "national security." In contrast, the environmental focus of the Roosevelt administration, as illustrated by continued investment in natural resource projects through the CCC and the Work Progress Administration, FDR's efforts to expand the National Park System to protect valued resources, and his ongoing commitment to private land stewardship through his continued commitment to the Soil Conservation Service, was consistent and substantial.

This essay explores evidence of FDR's early interest in and commitment to conservation, illustrated by the leadership provided during his tenure in the New York State Senate and as governor. It next discusses the conservation initiatives launched by FDR during his first 100 days in office. These initiatives had and continue to have a profound effect on the conservation of public and private land in the United States. The creative approaches taken by the FDR administration to address the problems of employment, to curb the excesses of agricultural production, and to promote public land protection continue to serve as the template for modern conservation and environmental policies.

As further evidence of the FDR environmental legacy, this essay highlights some of the key conservation appointments of the Roosevelt administration and the roles that each played in affecting conservation policy and environmental protection. Clearly, the leadership of Ickes, Wallace, Bennett, and Darling are reflective of the conservation-minded philosophy that was emblematic of FDR.

Finally, we discuss several of the significant accomplishments of FDR's tenure that serve to illustrate his personal commitment to environmental protection. For, it was in his creative and aggressive approach to resource conservation and environmental protection that FDR's environmental leadership becomes clear.

PLANTING THE SEEDS OF CONSERVATION LEADERSHIP

From an early age, Roosevelt exhibited an interest in the outdoors and in conservation. His attachment to the land and his interest in trees, agriculture, and the natural world are clearly rooted in his experiences here at his childhood home of Hyde Park. As observed by Ted Morgan (1985) in his biography of FDR,

> Above and beyond its physical setting, Hyde Park provided FDR with a set of values that were necessarily in the mainstream of American life. Hyde Park stood for the rooted versus the uprooted, for the rural versus the urban, for the sedentary versus the migrant, for he the old family versus the new arrival, for eastern gentility versus western crudeness, for tradition versus change. From these values, he took what he needed to project an image of himself, more mythical than real, as a farmer and country bumpkin and neighborly fellow who had a deep and abiding interest in the potato crop and apple barrels.[3]

It was, in fact, at Hyde Park that Roosevelt developed a passion for trees. His interest was such that he knew the species and qualities of each tree, their use, and their growth habits (i.e., distinguishing between those species that grew in the shade and those that required full sunlight). Over the years, it is reported that FDR planted some 220,000 trees at Hyde Park. As his letters illustrate, FDR's interest in and love of trees continued throughout his life. A letter to FDR from Nelson Brown of the New York College of Forestry at the beginning of his second term in office offered a report to the president on the planting of 26,000 trees on his "place," specifying, in great detail, the species and spacing of the trees planted.[4]

Similarly, FDR exhibited a passion for agriculture and identified himself with farmers and as a farmer in many of his speeches and correspondence.[5] Although it was an interest in which he developed considerable expertise, as noted by Morgan (1985),

> FDR was a farmer the way Einstein was a violinist. He knew quite a bit about it, but he did it as a hobby.

Among FDR's early passions was the collection of birds from the area surrounding Hyde Park in Dutchess County. It is reported that at the age of 11, Roosevelt was given a gun and set out, a la James J. Audubon, to collect one of every species of bird in the county. For two years, Roosevelt embarked on his quest and at the age of 14 presented the New York Museum of Natural History with a rare pine grosbeak skin. In return, FDR was offered an associate membership in the American Ornithologists' Union by museum curator F.M. Chapman.[6]

As noted by Davis (1985) in his biography of FDR during his New York years, it was James Roosevelt, FDR's father who helped to instill in FDR a love for the outdoors (which Davis referred to as "the manly virtues").

> He was an enthusiastic horseman; he actively supervised the farming of his acres, the management of his woods and herds of fine dairy cows; he loved

iceboating, hiking, tennis, sailing on the Hudson—even more sailing in the Bay of Fundy off the island of Campobello, where he established his summer vacation home when Franklin was in the second year of life. He involved his son in all these activities.[7]

Morgan (1985) observed that FDR had an affinity for history and a feeling that the American past, "was a manual of personal instruction." Morgan noted similarly, that,

> As he made the past his own, he made the land his own, beginning with Hyde Park, understanding what it meant to own a large estate, to farm it, [and] to plant trees. . . . [T]hrough the filter of the agrarian mystique, the idea that he was a farmer and that all farmers were kin, he could extend his sense of the land to the entire country.

FDR's passion for the land, his interest in agriculture and in trees carried forward as he entered the New York State Senate in 1911. Representing a predominantly rural and agricultural region of the state, Roosevelt soon found himself chairman of the Forest, Fish, and Game Committee of the Senate.

As evidence of his interest in conservation affairs, the story is told by Morgan (1985) of the efforts of Frances Perkins to lobby FDR for his support for a 54-hour workweek for women and children. When finally confronted by Ms. Perkins to secure his support for the bill on humanitarian grounds, it is reported that Roosevelt replied that he had "more important things" with which to deal. The "more important things" that reportedly preoccupied Roosevelt were bills to lower the fee for shad fishermen, to regulate the rough grouse season and a fight over a bill that permitted certain birds to be caught for their plumage and that extended the duck hunting season on Long Island.[8] A risky response for a state senator whose district included Vassar College.

Among the conservation measures being championed by Roosevelt at the time was the Roosevelt–Jones bill (Senate Bill No. 92) that called for state inspection of private forests, compulsory reforestation of denuded watershed lands, and state regulation of timber harvest to ensure permanent cover. Opposition from the timber lobby was strong. However, through two days of committee hearings, Roosevelt succeeded in bringing the matter to the attention of a number of state senators by inviting Gifford Pinchot, first chief of the U.S. Forest Service and a close associate of Teddy Roosevelt, to testify before the committee. During his presentation, Pinchot produced two pictures depicting the "before and after" conditions of a Chinese town, which Pinchot alleged had been left to ruin as a result of destructive logging practices. The display had such an effect on Roosevelt that he referred to it in a subsequent speech before the Troy, New York, People's Forum on March 3, 1912, remarks made more interesting by Roosevelt's reference to the distinction between the liberty of the individual and that of the community,

> Every man 500 years ago did as he pleased with his own property. He cut the trees without affording a chance for reproduction and he thereby parched the ground, dried up the streams and ruined the valley and the sad part of it is that there are to-day men of the State who for the sake of lining their pockets during their own lifetime are willing to cause the same thing that happened in China. . . . They care not what happens after they are gone and I will go even further and say that they care not what happens even to their neighbors, to the community as a whole, during their own lifetime. The opponents of Conservation . . . are merely opponents of the liberty of the community. . . .[9]

This reference to the welfare of the community, it seems, is a concept akin to the "Tragedy of the Commons" concern, which is, today, a cornerstone of conservation philosophy. In so advocating, Roosevelt also began to frame a stronger role for government in directing the economy, using his knowledge of conservation as a foundation for this philosophy,

> As with the conservation of natural resources so also is it bound to become with the production of food supply. The two go hand in hand, so much so that if we can prophesy today, that the state (in other words the people as a whole) will shortly tell a man how many trees he must cut, then why can we not, without being called radical, predict that the state will compel every farmer to till his land or raise beef or horses. After all, if I own a farm of a hundred acres and let it lie waste and overgrown, I am just as much a destroyer of the liberty of the community—and by liberty we mean happiness and prosperity—as the strong man who stands idle on the corner, refusing to work, a destroyer of his neighbor's happiness, prosperity and liberty.[10]

Senate Bill No. 92 was never reported from Roosevelt's committee, but the corresponding Assembly bill was passed and became law on April 12, 1911. Roosevelt introduced three other bills in this session of the New York State Senate relating to forestry, encouraging reforestation of private forest lands, the restoration of trees on denuded watershed lands, and promoting the control of forest fires.[11]

As Roosevelt took the helm as governor of New York in January 1929, he continued to demonstrate an interest in agriculture and conservation, particularly related to the husbandry of trees. He paid particular interest to the role of farm woodlots in providing timber and opportunities for profitable returns for farm owners, a subject he addressed early in his tenure at the annual meeting of the New York State Forestry Association[12] and then again, later in an article printed in *The Country Home* in June 1930.[13] It was in the context of this article that Roosevelt raised the notion of creating a "Grandchildren's Trust" as a means of encouraging investments in the reforestation of abandoned lands.

It was in the context of this article that FDR referenced, again, his experience in the State Legislature as chairman of the Forests, Fish, and Game and the two pictures used by Gifford Pinchot to illustrate the importance of conservation and reforestation to the well-being of rural communities and, in fact, the nation. He stated,

> One need not be an alarmist to foresee that, without intelligent conservation measures, long before half a millennium passes some such contrasting pictures might be possible in our own United States.[14]

It was clear from his statements as governor, that FDR saw a larger vision of conservation for the State of New York, a vision that most certainly affected the conservation strategies he adopted in later years as president. This conservation vision was articulated in FDR's message to the legislature in January 1931, in the context of a proposal to adopt a "land policy for the state." This policy, ostensibly, would seek to promote the survey of all rural lands so as to determine their highest and best use. FDR emphasized that of the 30 million acres of land in New York State, 27 were rural and nonindustrial. Of these 18 million were dedicated to farming.

Roosevelt's proposal was to determine which lands should best be used for farming and which would better be used for forestry. The survey, as proposed by FDR, would provide information on the best use of these lands as a basis for planning future state and local developments and aiding in rural economic strategies. In this context, FDR demonstrated an appreciation for the conservation value of these lands and for the need to reforest abandoned acres. Most importantly, Roosevelt demonstrated an appreciation for the use of science in making such land-use determinations and for connecting the prosperity of rural communities to the productivity of the land. It is in these comments that FDR's greater vision for conservation and his appreciation for the connection between the ecological and economic well-being of people and communities begins to emerge.[15]

Not long after his address to the State Legislature regarding a state land survey, FDR addressed a group assembled at the College of Agriculture, Cornell University, to discuss his conservation agenda for the state.[16] In particular, Governor Roosevelt discussed the future of the Adirondack Park and the Catskill Mountains, and pending legislation and constitutional amendments affecting each. The legislation pending would extend the existing boundary of the Park, an area in which development and the harvesting of timber from state-owned lands was restricted. Of the two constitutional amendments, one would have required a permanent reforestation program for the state and would provide annual appropriations for that purpose.

In the context of his remarks, Roosevelt made clear his support for extending the boundary of the Park and for the creation of a permanent program of reforestation for the state, a position he reiterated in a speech to the citizens of the State of New York later that fall.[17]

Another amendment to the state constitution favored by Roosevelt would authorize $19,000,000 for the purchase of submarginal land upon which trees would be planted. The funds would be expended over an 11-year period.

The measure had the strong support of conservation and sportsmen's organizations in the state as well as the bipartisan support of legislative leaders. Yet, to Roosevelt's surprise, at the last moment, the measure was strongly opposed by Al Smith, who had become increasingly alienated from Roosevelt. Smith called the measure "Socialistic" and claimed that it would put the state in competition with private enterprise by permitting the state to cut and sell trees grown on the purchased acres.[18]

Roosevelt responded to the challenge aggressively. Every Democratic worker in the state received a letter, drafted by Roosevelt, urging support for the amendment. Roosevelt made a radio address urging the amendment's support and criticizing Smith's opposition as politically motivated. In the end, the amendment was passed by a wide margin, providing $20,000,000 over a 15-year period for the acquisition of land and the planting of trees.[19]

The notion of converting marginal and abandoned agricultural lands to timberland through reforestation remained a continued focus of FDR. In his acceptance speech for the Democratic presidential nomination in Chicago on July 2, 1932, FDR linked the prospect of converting millions of acres of marginal lands to timberlands through reforestation and the prospect of immediate employment.

He argued the need for a "definite land policy," without which, "we face a future of soil erosion and timber famine." He continued,

> It is clear that economic foresight and immediate employment march hand in hand in the call for the reforestation of these vast areas. In so doing, employment can be given to a million men.[20]

Later, FDR expanded the notion of employment opportunities associated with "reforestation" to cover, in his words, "all aspects of the protection, conservation and enlargement of our forests. . . ." He went on to note,

> In the vast national forests there is opportunity and need for a greatly increased program of improvement. This would give work to many thousands of men

during the present emergency. One of the prime needs is for road and trail building for fire protection and funds for this purpose would, in my judgment, be a size expenditure to be classed as dividend-paying capital investments. There is also in these forests the opportunity for tree planting and improvement cuttings.[21]

Thus, as FDR prepared to enter the White House in the fall of 1932, his interest in trees and conservation, and his commitment to the reforestation of marginal, abandoned, and cutover lands had grown into a proposal to employ millions of unemployed Americans. In response to criticism and comment, his notion of "reforestation" had expanded to include all manner of forest improvements and conservation measures intended to protect forests from fire, to expand access through road construction and to improve recreation opportunities through the construction of trails and other recreation facilities.

It is interesting to note, that as FDR entered the White House in January 1933, he received a letter from Governor Gifford Pinchot, governor of Pennsylvania, who urged FDR's continued focus on forestry and the opportunities associated with the purchase of additional private forest lands to be converted to national forests. Pinchot stated,

> . . . Work relief is better than direct relief, and useful labor paid for from relief funds is pure gain. By utilizing the unemployed highly necessary and productive improvements could be made in the forests thus acquired at substantially no cost to the public. The major fields of work include planting, thinnings, release cuttings, the removal of highly inflammable snags and windfalls, a large scale attack on serious insect epidemics, the control of erosion, the construction of roads, trails, and telephone lines, and the development of camp sites and other recreational facilities. All this would supply large numbers of men with highly useful work. . . .
>
> [A]s I see it there is no single domestic step that can be taken that will mean so much to the future of the United States as this one. . . .[22]

PRESIDENTIAL INITIATIVES TO PROMOTE CONSERVATION AND ENVIRONMENTAL PROTECTION—THE FIRST ONE HUNDRED DAYS

It is clear that FDR set a high standard for initiative and presidential leadership in the first 100 days of his first term in office. While much has been written about the successful legislative and administrative initiatives of FDR during this period, few observers have taken adequate note of the many other important actions initiated by FDR's administration during this time period to promote conservation and environmental protection. While designed primarily to stoke the national economy and put Americans back to work, these initiatives also had important and lasting conservation and environmental protection implications.

THE AGRICULTURAL ADJUSTMENT ACT (AAA)

Congress acted quickly at the behest of the Roosevelt administration to pass legislation to address the despair felt by many farm families and rural communities in the early 1930s. One element of the farm program established was an effort to address crop surpluses by limiting production. In so doing, the new farm policy sought to achieve dual goals—to improve commodity prices through supply controls and to avert the further degradation of soil resources brought about by excessive soil erosion.

By paying farmers to idle lands and, thus, curb the production of commodities already in excess, a strategy was established that has remained a cornerstone of agricultural policy to this day. The popular and highly effective Conservation Reserve Program (CRP) that, today, idles nearly 45 million acres of highly erodible croplands in the United States has, as a fundamental goal, the reduction of soil erosion through land retirement. Modifications of this program have enhanced its ability to protect water quality and enhance wildlife habitat. As a result, the CRP has the strong backing of a broad coalition of commodity, conservation, wildlife, and environmental groups.

It is interesting to note that one feature of the AAA, the creation of state and local committees of farmers empowered to implement acreage controls at the local level, has worked both to benefit and disadvantage subsequent efforts to further conservation and environmental protection. In their establishment, the county committees were intended to serve as a "built-in protection against overcentralization . . . [and] as a buffer between [farmers] and Washington." In this, the county committee system has been extremely successful. However, while providing this buffer, some critics of the committee system have argued that it has thwarted efforts to promote new conservation measures or policies.

CIVILIAN CONSERVATION CORPS

Clearly, the CCC had a tremendous impact on the nation's natural resources and a beneficial effect upon the thousands of young adults who participated in the effort. As a precursor to the CCC, Roosevelt had employed over 10,000 men from the relief rolls of New York in the reforestation of cutover forests and abandoned agricultural lands.

Miller said of the CCC that,

> Of all the legislation of the Hundred Days, the CCC was the closest to Roosevelt's own heart. Expressing his lifelong interest in forests and conservation, it was designed to put two hundred fifty thousand unemployed young men to work in the forests and national parks on reforestation and flood-control projects.[23]

The nature of the work undertaken and the breadth of its impact were substantial. From 1933 to 1942, the CCC planted millions of trees, constructed thousands of miles of roads and trails, laid telephone lines, fought forest fires, built erosion control structures, and improved range, wildlife, and fish resources. In fact, in June 1935 the *New Republic* dubbed the CCC "Roosevelt's Tree Army."[24] In aggregate, over 3 million individuals participated in the program. During its eight years, the CCC created up to 800 state parks, 46,000 campgrounds, 3,116 fire towers, restored over 3,980 historic structures, laid 97,000 miles of road, restored 80,000,000 acres of farmland and planted well over 3 billion trees.[25] The estimated value of this work was $1,750,000,000 in 1942.

The work of the CCC was not limited to public lands and public works projects. As Roosevelt relocated the Soil Erosion Service to the Department of Agriculture (and it was renamed by Soil Conservation Service (SCS) in accordance with the Soil Conservation Act of 1935), the SCS took over more than 150 CCC camps previously under the general supervision of the Forest Service.[26] The work of the CCC was implemented under the guidance of the SCS and Hugh Hammond Bennett focused on the development of demonstration projects to illustrate improved cropping methods for reducing soil erosion, the collection of native seeds and the propagation of grasses and trees for erosion control, and the construction of terraces, ditches, and ponds to curb soil loss and improve water quality.

As emphasized by Helms (1992), the benefit of these projects and the SCS's involvement in the CCC was substantial beyond their immediate value in curbing soil erosion. Recognizing that, at the time, conservation was essentially an experiment, Helms noted,

> The work of the CCC crews was valuable to Bennett in proving the validity of his ideas about the benefits of concentrated conservation treatment of an entire watershed. The large-scale approach also permitted experimentation. Few of the conservationists' techniques were new, but the process of fitting them together was. The work led to the refinement and improvement of conservation measures still used today.[27]

The contributions of the CCC, as significant as they were for improving natural resource and environmental conditions from 1933 through 1942, had an even greater and more lasting impact. Nationally, 90 percent of all CCC enrollees took classes and 100,000 men were taught to read and write. Five thousand earned high-school diplomas.[28]

The CCC experiment, as it has been referred, has influenced the development of the Peace Corps, the Job Corps, Youth Conservation Corps, and Americorp. It has demonstrated the value of investing in natural resource and environmental protection as a means of providing education, income, and a character-building experience for its participants. As evidence of its popularity, in 1936, Republican presidential nominee Alfred M. Landon endorsed the CCC and the *Detroit News* noted that, although expensive, "the prompt and unmistakable dividends it has paid, both in valuable work accomplished in the nation's forests and in the physical and moral benefits accruing to the young men who have enlisted made it a real investment in the National well-being." In fact, a July 1936 poll found that 82 percent of Americans supported the CCC.[29]

THE PUBLIC WORKS ADMINISTRATION

The Public Works Administration (PWA), operated under the direction of Secretary of the Interior Ickes, and established as a part of Title II of the National Industrial Recovery Act, sought to help jumpstart the economy through an infusion of federal funds in public works projects. The bill authorized $3.3 billion for projects and funded the completion of such New Deal monuments as Hoover, Boulder, Grand Coulee, and Bonneville Dams.

Of course, the PWA did not just build dams. It constructed schools, hospitals, municipal water systems, sewer lines and sewage disposal plants, and local street and highway improvements. Other projects of the PWA included the Timberline Lodge at the top of Mount Hood in Oregon and national parks projects such as the Blue Ridge Parkway, restoration of the Lee Mansion in Arlington National Cemetery, and administration buildings in Shenandoah, Great Smokey Mountains, and Olympic National Parks.[30]

When Congress passed the Emergency Relief Appropriations Act in April 1935, providing $4.8 billion for work programs, it was the largest peacetime appropriation in American history.[31] Many of the projects built by the PWA remain important components of the nation's outdoor recreation infrastructure today.

THE TENNESSEE VALLEY AUTHORITY

When enacted in April 1933, the legislation authorizing creation of the Tennessee Valley Authority (TVA) created a great experiment in regional planning. Embodying the concepts of public power and conservation, the project sought to improve and protect the resources and communities in the watershed of the Tennessee River, comprising 293,000 acres in seven states. The act provided for the only "unified source-to-mouth treatment of a great river and its tributaries" through flood control by a series of dams and reservoirs; development of navigation; generation of electricity; use of marginal lands; reforestation of all lands in the basin suitable for reforestation; and economic and social well-being of the people living in the river basin.[32]

Excessive timber cutting coal mining, abusive farming, and mining had devastated the watershed of the Tennessee River. As the productivity had been stripped from the land, so, too, had opportunities for economic growth and rural prosperity.

As acknowledged by TVA's first director, David Lilienthal,

> Saving the soil is the first step—to keep the soil from washing away; to save its fertility from being hopelessly exhausted. For the soil is the nation's basic resource.[33]

The TVA established a forestry division and set about combating soil loss through a massive reforestation effort. Joining forces with the CCC—about 20 camps and 6,000 participants were located in the Tennessee Valley—they established check dams, built diversion ditches, improved existing forest stands, and established new ones. At the same time, farmers in the valley were given assistance in restoring the productivity of their land through improved cropping and the use of fertilizer to stimulate new crop growth. Through demonstration projects, a device employed by the SCS through the work of the CCC, farmers and forest landowners were educated in new methods of conservation and resource restoration.

At the outset, TVA Director Lilienthal called the program of flood control, land recovery, power development, and human betterment a "seamless web."[34] In reflecting on the role of TVA in the valley, Lilienthal is reported to have said,

> ... TVA would provide the tools of opportunity—flood control, malaria control, navigation on the river, low cost power, test-demonstration farming to show how our soils could be returned to fertility, a fertility lost through land erosion, another wayward child of a one-crop system. He told us the river would no longer defeat man, but would become the servant of man. "What you do with these tools," he said, "is up to you."[35]

CONSERVATION LEADERS IN THE ROOSEVELT ADMINISTRATION

Even less credit is given to FDR for the wisdom and foresight he demonstrated in the selection of individuals who would prove themselves to be conservation leaders during the Roosevelt administration.

HAROLD ICKES AND THE DEPARTMENT OF THE INTERIOR

As with any administration, the work of environmental protection and conservation leadership is done in the departments and agencies entrusted with that role. So, too, was true with the Roosevelt administration. And, as in other administrations, the primary work of conservation resided in the agencies in the Department of the Interior and the Department of Agriculture.

For leadership at the Department of the Interior, Roosevelt chose Harold Ickes. Ickes, an attorney and advocate for civil rights and social justice from Chicago had been a political operative for the FDR campaign in the Midwest. Ickes was asked by the campaign to organize the Midwest Independent-Republican vote for their candidate,[36] although his decision to support the Democratic candidate for president was not well received by his wife who was running for a state legislative position as a Republican. Ickes strategized that his support for Roosevelt might lead to an opportunity for appointment to a position in the Department of the Interior, where he might work on behalf of the welfare of the Indians, a concern of his wife, Anna. Of course, as history and opportunity would have it, after a number of individuals had turned down FDR's offers to head the Department, Roosevelt turned to Ickes to serve as secretary, as he needed "a man from the West, an honest man, who could stand up to the interests."[37]

Ickes set about righting many of the wrongs that had afflicted the Department of the Interior and had resulted in a public perception that the Department was beset by corruption and mismanagement. Early on, he articulated his view of the Department's role: "Our primary concern is the protection and enlargement of life and the conservation of natural resources." He set about reversing the perception that the Department was "almost irredeemably corrupt" a view affirmed by the conviction in 1927 of former Interior Secretary Albert Fall for accepting bribes to secure the leasing of two oil reserves in Wyoming for commercial interests (popularly known as the Teapot Dome scandal).[38]

Similarly, Ickes sought to promote greater social justice within the Department by an early directive to end racial discrimination in the Department's rest rooms and cafeteria.[39]

Roosevelt illustrated great faith in Ickes in appointing him to administer the work of the newly created PWA. And, this clearly worked to the benefit of the agencies under Ickes's domain. The National Park Service was able to accomplish much-needed rehabilitative work under the PWA and to expand its mission through the creation of The Blue Ridge, Skyline Drive, and Natchez Trace scenic parkways. In addition, the Jefferson Memorial was constructed near the mall and the National Park System dramatically expanded. As noted by Clarke (1996) the 1930s was one of the few periods in National

Park Service history when it has had money to spend on something other than absolute necessities such as operations and maintenance.[40]

Other notable conservation leaders in the Department of the Interior included Horace L. Albright, Director of the National Park System and Hugh Hammond Bennett, who directed Interior's Soil Erosion Service prior to its' transfer to the U.S. Department of Agriculture and conversion to the SCS.

HENRY WALLACE AND THE DEPARTMENT OF AGRICULTURE

In selecting Henry Wallace to run the Department of Agriculture, FDR chose a man who was respected for his honesty and integrity and committed to rapid action to address the plight of farmers and rural communities suffering from the downturn in the U.S. economy. While it is true that Harold Ickes is most often viewed as the lead conservationist in the FDR cabinet, it is important to recognize that most of the conservation-minded agencies of the federal government at that time, resided in the Department of Agriculture.

The Forest Service, transferred from the Department of the Interior in 1905, was USDA's most visible conservation organization. In fact, it was Ickes's obsession with returning the Forest Service to the Department of the Interior, that clouded their working relationship and eventually led to a rift between Ickes and Wallace.

The Bureau of Biological Survey, precursor to the U.S. Fish and Wildlife Service, was another of the conservation agencies under Wallace's domain. Led by conservationist and Pulitzer Prize winning cartoonist Jay N. "Ding" Darling, the Bureau was an aggressive advocate for the protection of game birds, waterfowl, and their habitats.

To illustrate this point, on August 1, 1935, Secretary Wallace and FDR signed regulations for the hunting of waterfowl that would reduce bag limits, shorten the hunting season, prohibit the use of live decoys, and abolish the shooting of waterfowl over baited waters. This action was taken despite strong opposition from Roosevelt's wealthiest friends and enemies.[41]

Through subsequent reorganization of the federal government's conservation agencies, the Bureau of Biological Survey was subsequently transferred to the Department of the Interior, where, in combination with the Bureau of Fisheries (which had been transferred from the Commerce Department) it became the U.S. Fish and Wildlife Service. It was in this same transfer that the Soil Erosion Service and its head, Hugh Hammond Bennett, were sent from the Interior Department to USDA. This marked a return for Bennett to his roots in the Department of Agriculture, and set the foundation for the creation of the SCS that became the lead agency in combating soil loss on the nation's private agricultural lands.

OTHER COMPONENTS OF FDR'S
ENVIRONMENTAL LEGACY—CONSTRUCTING
A NEW FRAMEWORK FOR CONSERVATION

Through efforts to reorganize the conservation agencies of the federal government, to promote new legislative authorities for conservation, to promote expansion of the national parks and forests, and to create new means to enhance the nation's ability to promote the conservation of its natural resources, the Roosevelt administration demonstrated a strong commitment to conservation and environmental protection.

As noted above, both Henry Wallace and Harold Ickes expressed interest in consolidating the agencies and the functions of government that could protect the conservation and protection of the nation's natural resources. Ickes was particularly interested in transforming the Department of the Interior into "a genuine Department of Conservation" by expanding and redirecting the national park system, restoring the Forest Service to the Department of the Interior, and "collect[ing] the other federal resource-managing agencies from the departments of his cabinet colleagues."[42] While, to Ickes's dismay, Wallace had in mind a different restructuring—namely to move the national parks under the jurisdiction of the Department of Agriculture—no such scheme ever came about. Nevertheless, subsequent moves to establish the SCS, the U.S. Fish and Wildlife Service, and to enhance Interior's ability to manage the public lands further illustrated that conservation was a continuing priority of the Roosevelt administration.

Among Ickes's accomplishments was reorganization of the National Park Service. On June 10 and July 28, 1933, Roosevelt signed executive orders that consolidated all existing national parks, national monuments, military sites, cemeteries, and monuments in the nation's capital under the single authority of the Park Service. And, during the summer of 1933, over 50 separate areas previously administered by the War and Agriculture departments, such as the Gettysburg battlefield, were incorporated into the national park system, transforming the agency into a national entity responsible for the conservation of natural environments, and cultural and historical resources.[43] FDR designated 1934 "National Parks Year," and spent much of August of that year touring the western parks with his family and Secretary Ickes.[44]

With the help of Ickes and Wallace, during his tenure, FDR engaged in an aggressive effort to expand both the national parks and forests. During Ickes's tenure 47 new park areas were added to the national park system, including areas from the Everglades in Florida to the Olympic National Park in Washington. As noted by Clarke (1996),

[T]he [National Park] system was much larger, much more diverse, and better managed after Ickes' thirteen years in office. By any measure, it was an impressive record and a boon to the agency.

At the same time, FDR directed an expansion of the national forest system to acquire lands east of the Mississippi River. He sent a memo to Wallace in July 1933, and commented that he viewed the potential for additional national forest land purchases "with some interest."[45] Subsequently, some 22 million acres were added to the system through the acquisition of cutover and abandoned lands during the mid-1930s.

Not only were the national parks and forests systems expanded during FDR's tenure, but through the urging of Ding Darling, a system of game refuges—precursor to the National Wildlife Refuge System—was also established. With the passage of the Taylor Grazing Act, Congress moved to establish new authorities for the management of public grazing lands, to be administered by the Division of Grazing in the Department of the Interior. Darling recommended the establishment of numerous big-game refuges on the public lands, to be administered, instead, by the Bureau of the Biological Survey. FDR agreed, and, over the objections of Taylor, granted authority for determining how many head of cattle should be permitted to graze on the refuges to the Biological Survey.[46]

In fact, the Taylor Grazing Act that created the Grazing Service in the Department of the Interior was viewed as one of the earliest New Deal contributions to the conservation cause ending the Department's free or nearly free land-disposal policy.[47]

The Roosevelt administration also played a role in promoting the improved use of science in fish and wildlife management, largely through the initiative of Ding Darling. In Iowa, Darling had been instrumental in the establishment of the first Cooperative Wildlife Research Unit, housed at Iowa State College. The unit was staffed by a Ph.D. graduate of the University of Wisconsin and several graduate students and funded, in part, by Darling himself. As Darling became the head of the Bureau of Biological Survey, he immediately sought support for the establishment of a network of similar units and succeeded in securing nine. Of the 9 original units, 7 are part of a network of 43 such units currently operating in 40 states under the auspices of the Biological Resources Division of the U.S. Geological Survey.[48]

Darling also played an instrumental role in establishing the federal duck stamp that continues to provide a significant source of revenue for the purchase of wetlands and migratory bird habitat in the United States. Largely through his artistic efforts and political cartoons, Darling had highlighted for the nation, the effects of destructive farming practices and the Dust Bowl on the nation's waterfowl and their habitats. Although the Migratory Bird Conservation Act of 1929 had been enacted to help address the situation, the Congress had failed to create a mechanism to fund the needed work. Darling and FDR were instrumental in encouraging development and passage of the Migratory Bird Hunting Stamp Act in March 1934, which, through the sale of duck stamps to waterfowl hunters, provided a funding mechanism for habitat restoration and land acquisition. In fact, it was Darling who designed the first duck stamp issued under the Act.[49]

Roosevelt also played a personal role in promoting the conservation and restoration of wildlife. In 1935, Roosevelt wrote Forest Service Chief Silcox calling for a North American Wildlife Conference to bring together those interested in the restoration and conservation of wildlife resources. This was the first coming together of the broad range of wildlife groups engaged in conservation efforts across the nation. In his letter of confirmation and his charge to Silcox, Roosevelt noted,

> Conservation of our remaining natural resources through use is vital to the social and economic welfare of the nation. Despite real program in conservation of such resources as minerals, forage, forests, and the soil itself, much remains to be accomplished. This is particularly true with respect to the restoration of wildlife.[50]

One outcome of this conference was support for enactment of the Pittman–Robertson Act of 1937, which established a fund for state fish and wildlife programs from the proceeds of federal taxes on hunting.

In addition, on January 6, 1934, FDR appointed a special committee, composed of Thomas Beck, Ding Darling, and Aldo Leopold to "investigate the wildlife problem, and to draw up a national plan and policy to provide for the restoration and maintenance of this valuable renewable resource."[51] The report of the committee outlined a plan for restoring wildlife through land acquisition and habitat restoration activities that guided the acquisition of nearly 5 million acres of wildlife refuge lands.

Most fascinating was the role that FDR personally played in the creation and expansion of the Olympic National Park and Kings Canyon National Parks, expansion of the Grand Teton National Park, and further protection of Admiralty Island in southeast Alaska. These, and other actions taken to expand protections for notable scenic and ecological resources within the public domain, illustrate the breadth of FDR's interest in and commitment to environmental protection.

FDR's Environmental Leadership: Innovation, Experimentation, and a Commitment to Conservation

As evidenced by the breadth and impact of the accomplishments of the Roosevelt administration in conservation and environmental protection, FDR and his administration have left a lasting legacy on the American landscape.

Through innovative approaches to addressing the dual problems of environmental degradation and economic hardship at the beginning of the 1930s, FDR demonstrated a capacity to solve human problems by investing in natural resources and the environment. The CCC, a model for subsequent efforts to invest in people and the environment through government funded environmental projects, illustrates, well, the creative approach that FDR took to dealing with the social problems of the day. The extensive investment made through the PWA in projects designed to both produce public power and check soil erosion and the risk of flooding serve to illustrate how human ingenuity could be harnessed to address both human and environmental needs. At the same time, other investments made by the PWA such as the expansion of recreation facilities and the construction of new trails, scenic highways, and the restoration of cultural and historic sites, demonstrated, clearly, that Roosevelt understood the value to the national "psychology" of investing in tangible and lasting improvements in the environment.

FDR's willingness to experiment with new approaches to promote environmental protection provided a new paradigm for the role that government could play in improving the lives of farmers and the residents of rural communities. The TVA best exemplified this willingness to experiment and to test the boundaries of public (and congressional) acceptance of new roles for government in aiding the people it serves. In the case of the TVA, not only did the government play a new and expansive role in affecting conservation practices on the ground, but it did so on a scale that had never before been imagined. The establishment of a new agency—the SCS—dedicated to the conservation of soil and natural resources on private lands, also demonstrated FDR's willingness to try new approaches to improving the nation's natural resources. In fact, this remains the model for local partnerships to promote soil and water conservation across the nation.

FDR's contributions to environmental protection and conservation affected the entirety of the American landscape. While the work of the SCS focused on the conservation of soil and water resources on private lands, the work of the Forest Service, the Bureau of the Biological Survey, and the National Park Service addressed the needs of public lands and resources. Innovations such as the Taylor Grazing Act further improved the management and protection of public land resources. But, what was unique for the time and an important foundation for the conservation work before us today, was that the conservation initiatives of FDR's administration benefited both public and private lands. Acknowledging, as we do today, that the environmental challenges of the nation require a regional or landscape perspective, the approach of the Roosevelt administration was truly novel for its time.

Recognizing that the welfare of man and the well-being of the land are inextricably linked, was also a novel concept at the time. As noted by Munzer in her study of the TVA,

> The idea that man and his environment are part of a "seamless web" had taken a long time to mature and a longer time still to find acceptance.[52]

This concept is fundamental to understanding the notion of sustainability which, today, serves as the foundation for much of the conservation and environmental work we do. FDR and his administration understood this inextricable connection between social, economic, and ecological well-being and used it as a template for the natural resource and environmental policy they pursued. The success of this strategy—of this great experiment in sustainability—is a lesson worth reviewing in the current political climate.

Finally, FDR's environmental leadership, like that of many environmental leaders, was borne of a personal commitment to and passion for nature and the outdoors. The influences of his father and of his experiences at Hyde Park leading to his interest in and love for trees clearly planted in FDR the seeds of an environmental leader. These environmental roots began to show themselves early in FDR. As a state legislator and then, later, as governor, FDR demonstrated his passion for trees and protecting the environment. His association with mentors like Gifford Pinchot, and, perhaps, the conservation legacy of his distant cousin, Teddy, had their influences as well. Whatever its origin, FDR's passion for conservation and environmental protection were a compass for many of the policy decisions he made affecting the environment and the nation's natural resources throughout his professional career.

FDR's environmental legacy and the evidence of his environmental leadership is as apparent today as it was a half century ago. His leadership and the work of his administration changed the American landscape at a time when environmental degradation was destroying the productivity of our land and, with it, of the nation. The investments he made in parks and forests created the roads, trails, and campgrounds we enjoy today. The federal agencies he created and the legislative authorities he signed into law still provide the institutional and legal framework for much of the conservation work we do today. And, the concepts and principles of conservation FDR and his administration developed and implemented are, in fact, fundamental to how we view contemporary environmental challenges and frame their solutions.

NOTES

1. Burns, James MacGregor, 1978. *Leadership* 1.
2. Report of the National Energy Policy Development Group, the "National Energy Policy report" was delivered to the President in May 2001. The report

emphasized, "The first step toward a sound international energy policy is to use our own capability to produce, process, and transport the energy resources we need in an efficient and environmentally sustainable manner."

3. Burns (1978) emphasizes the important role of childhood experiences in affecting the development of values inherent in leadership.

4. Letter from Nelson C. Brown to President Roosevelt, dated May 22, 1937. In Nixon (1957) *Franklin D. Roosevelt and Conservation 1911–1945, Volume II,* 65.

5. Morgan (1985), 38.

6. Ibid., 45.

7. Davis (1985), 5–6.

8. Morgan (1985), 127.

9. Excerpt from remarks provided by FDR before the Troy, New York People's Forum. In Nixon (1957), Vol. I, 18–19.

10. Miller (1983), 90.

11. Nixon (1957) Vol. I footnote, 20–21.

12. Remarks of Governor Franklin D. Roosevelt's Address to the New York State Forestry Association, Seventeenth Annual Meeting, Albany, NY, February 27, 1929. In Nixon (1957), Vol. I, 69–71.

13. Nixon (1957), Vol. I, 71–76.

14. Ibid., Vol. I, 72.

15. The Land Survey of the State of New York, as outlined by Governor Franklin D. Roosevelt in his message to the legislature, Albany, January 26, 1931. In Nixon (1957), Vol. I, 77–81.

16. Speech by Roosevelt, Farm and Home Wee, Cornell University, Ithaca, New York, February 13, 1931. In Nixon (1957), Vol. I, 81–84.

17. Speech by Roosevelt to the Citizens of the State of New York, October 26, 1931. In Nixon (1957), Vol. I, 94–98.

18. Davis (1985), Vol. I, 243–46.

19. See Nixon (1957), Vol. I, 110.

20. From Roosevelt's acceptance speech, Democratic National Convention, Chicago, July 2, 1932. In Nixon (1957), Vol. I, 108.

21. From a letter from Roosevelt to Miller Freeman, dated August 6, 1932. In Nixon (1957), Vol. I, 123.

22. From a letter to Roosevelt from Governor Gifford Pinchot of Pennsylvania, January 20, 1933. In Nixon (1957), Vol. I, 129–32.

23. Miller (1983), 315.

24. From http://www.michiganhistory.org/museum/techstuf/depressn/treearmy/html.

25. From http://www.geocities.com/cccorps/create/html.

26. Helms (1992), 47.

27. Ibid., 49.

28. From http://www.michiganhistory.org/museum/techstuf/depressn/treearmy/html.

29. Ibid.

30. From http://www.cr.nps.gov/history/online_books/unrau-willis/adhi3.htm.

31. Morgan (1985), 417.

32. Munzer (1969), 69.
33. Ibid., 148–49.
34. Ibid., 164.
35. Ibid., 165.
36. White and Maze (1985), 96.
37. Ibid., 98.
38. Ibid., 104.
39. Ibid., 105.
40. Clarke (1996), 161.
41. Brant (1988), 50–51.
42. Clarke (1996), 107.
43. Ibid., 111.
44. As noted by Clarke (1996), many observers believed that FDR's reorganization was the single most important event in the history of the National Park Service.
45. Roosevelt to Henry A. Wallace, July 11, 1933. In Nixon (1957), Vol. I, 190.
46. Brant (1988), 48.
47. Clarke (1996), 48.
48. See "History of the Iowa CFWRU" at http://www.ag.iastate.edu/centers/cfwru/history2.htm.
49. From http://duckstamps.fws.gov/history.html.
50. Roosevelt to Ferdinand A. Silcox, Chief, Forest Service, December 20, 1935. In Nixon (1957), 462.
51. From H.P. Sheldon, Chief, Division of Public Relations, Bureau of Biological Survey, to William D. Hassett, Special Assistant to Stephen T. Early, September 26, 1936. In Nixon (1957), 556.
52. Munzer (1969), 169.

REFERENCES

Brant, Irving. 1988. *Adventures in Conservation with Franklin D. Roosevelt.* Flagstaff, Arizona: Northland Publishing Company.

Clarke, Jeanne Nienaber. 1996. *Roosevelt's Warrior: Harold L. Ickes and the New Deal.* Baltimore, Maryland: The Johns Hopkins University Press.

Creese, Walter L. 1990. *TVA's Public Planning: The Vision, The Reality.* Knoxville, Tennessee: The University of Tennessee Press.

Daynes, Byron W., William D. Pederson, and Michael P. Riccards, eds. 1998. *The New Deal and Public Policy.* New York: St. Martin's Press.

Davis, Kenneth S. 1985. *FDR, The New York Years, 1928–1933.* New York: Random House.

Helms, Douglas. 1992. *Readings in the History of the Soil Conservation Service.* U.S. Department of Agriculture, Soil Conservation Service, Economics and Social Sciences Division, NHQ. Historical Notes Number 1. Washington, DC.

Munzer, Marthaq E. 1969. *Valley of Vision: The TVA Years.* New York: Alfred A. Knopf.

Miller, Nathan. 1983. *FDR: An Intimate History.* New York: Doubleday & Company, Inc.

Morgan, Ted. 1985. *FDR: A Biography.* New York: Simon and Schuster.

Nixon, Edgar B. 1957. *Franklin Roosevelt & Conservation: 1911–1945, Volume One.* Washington, DC: Government Printing Office.

Nixon, Edgar B. 1957. *Franklin Roosevelt & Conservation: 1911–1945, Volume Two.* Washington, DC: Government Printing Office.

Schapsmeier, Edward L. and Frederick H. Schapsmeier. 1968. *Henry A. Wallace of Iowa: The Agrarian Years, 1910–1940.* Ames, Iowa: The Iowa State University Press.

White, Graham and John Maze. 1985. *Harold Ickes of the New Deal.* Cambridge, Massachusetts: Harvard University Press.

White, Graham and John Maze. 1995. *Henry A. Wallace: His Search for a New World Order.* Chapel Hill, North Carolina: The University of North Carolina Press.

PART 4

A USABLE PAST

CHAPTER 10

RECOVERING FDR'S ENVIRONMENTAL LEGACY

RICHARD N.L. ANDREWS

INTRODUCTION

FRANKLIN D. ROOSEVELT (FDR) BECAME PRESIDENT during the worst conjunction of economic and environmental disasters in American history: the Great Depression and near-collapse of the American economy, and the Dust Bowl and devastating floods of the 1930s (Lowitt 1984, 35, 37). By the time of his death in office 12 years later, he left an environmental as well as an economic and administrative legacy that is arguably unmatched by any president before or since. His environmental legacy included three broad elements: a legacy of specific environmental policies, programs, and institutions; a legacy of values and principles for environmental leadership and management; and a legacy of environmental results, including the impacts both of his environmental initiatives and of many other initiatives intended to achieve other policy goals.

Roosevelt's New Deal put in place the principles of a powerful and distinctive vision of environmental policy: that the natural environment could be developed and managed in an integrated fashion for human benefit, that this could be done in ways that restored and conserved nature itself while also building a healthy economy and society, and that government leadership and planning, rather than just the invisible hand of the market, were necessary and effective instruments to accomplish this. This vision built heavily on the prior legacy of Theodore Roosevelt's Progressivism, but added distinctive innovations of FDR's New Deal agenda as well.

Roosevelt's environmental legacy is a record not solely of successes, but of pervasive influence on many of the policies, impacts, and controversies that have followed. It included important achievements and innovations, and new institutional structures, that have lasted far beyond his administration. It also included some unfulfilled hopes, and unanticipated and sometimes perverse impacts, as that legacy has been appropriated and adapted by others over the half century since. To many modern environmentalists, the environmental legacy of Roosevelt's New Deal is as much a source of current environmental problems as of solutions, as Daniel Tarlock notes pointedly in chapter 6 of this book. However, it is also a source of important achievements and lessons that have been too little recognized both by environmentalists and others.

To recover FDR's legacy today, then, has three purposes. One is to recall and reaffirm the content, magnitude, and historical importance of this legacy itself, whose details far too few today may remember. A second is to consider its impacts on subsequent events and issues, and the consequences— unanticipated as well as intended—which these initiatives helped to set in motion. Finally, a third purpose is to consider whether Roosevelt's initiatives may also offer any tools or lessons for addressing environmental policy issues today.

POLICIES, PROGRAMS, AND INSTITUTIONS

FDR's legacy of specific environmental initiatives included three broad elements: programs for environmental restoration from the devastating effects of soil erosion, deforestation, floods, and other damage; programs for environmental management and improvement more generally, and particularly for rural conservation and economic development; and major innovations in government institutions and policy incentives.[1]

ENVIRONMENTAL RESTORATION INITIATIVES

FDR in the 1930s faced a nation ravaged by years of drought and dust storms, deforestation, soil erosion, and devastating floods. One of his first priorities and most enduring successes was the restoration of deforested and eroded lands to healthy and productive ecosystems. No other president has contributed as much to the restoration of damaged environmental conditions and ecosystems as Roosevelt's New Deal.

The Civilian Conservation Corps (CCC), building on initiatives FDR had begun as governor of New York, employed over 2.5 million young men in restoration projects between 1933 and 1941. Overall, CCC workers planted an estimated 3 billion trees, stocked nearly a billion fish, built a million miles of roads and trails, and worked nationwide in soil erosion control, rangeland and habitat restoration, forest fire prevention, wildlife habitat improvement, drainage projects, and other activities (Schlesinger 1965, 337–40; Owen 1983, 128–45). By some accounts, more than half of all the trees ever planted in the United States, both publicly and privately, were planted by the CCC (Leuchtenburg 1963, 174 and notes cited).

For restoration of eroded agricultural lands, Roosevelt championed a program of voluntary, cooperative partnerships with farmers and local farm associations, coupled with technical and financial assistance, to promote retirement of marginal lands and introduction of soil-building practices on lands that remained in production. Under the Soil Conservation and Domestic Allotment Act of 1936, farmers who voluntarily restricted their acreage in production to allotted amounts, and used prescribed soil-building practices on those acres they did plant, would receive conservation payments from the federal government; payments were made only after the farmer's own county agricultural conservation association had made a performance check to verify them. These included such practices as crop rotation and contour plowing, terracing of slopes, planting of forest "shelterbelts" to reduce wind erosion, and watershed management. By the late 1930s such conservation practices had been put in place on nearly 53 million acres, involving partnerships with over 2,700 county agricultural conservation groups (*Public Papers* VII:89). To provide them technical assistance in these practices, FDR recreated the Agriculture Department's fledgling Soil Erosion Service as a Soil Conservation Service (SCS) with broader authority and a far larger staff and budget.

For restoration of the drought-devastated Great Plains, FDR's administration joined forces with western legislators to pass the Taylor Grazing Act of 1934, which designated 80 million acres of public lands as rangelands and provided that they be actively managed by a new Grazing Service under permits and user fees ("animal unit months," or AUMs). While the fees were arguably too low (due to the political influence of the ranching lobby), the Taylor Act for the first time provided both authority and dedicated revenues for federal restoration and active management of its vast western rangelands.

Finally, the New Deal made major contributions to environmental restoration through additions to the National Forest and National Park Systems. Most other presidents have protected such lands mainly by reserving them from the existing federal public domain. FDR, in contrast, targeted 75 million acres of submarginal, deforested, and eroded forest lands for repurchase from private owners under the authority of the 1911 Weeks Act, and for subsequent restoration as national forests and parks. These lands form many of today's eastern national forests and national parks, making the national park and forest systems a fully national rather than merely western heritage. Examples included the Great Smokies and Everglades National Parks (and Olympia in the Northwest), Cape Hatteras National Seashore, the Blue Ridge, Natchez Trace, and George Washington Parkways, as well as many eastern national forests. More than a million acres were also transferred to the states, and the Fulmer Act of 1935 helped the states to buy up and protect an additional 19 million acres of forest lands while they were available at Depression era prices.

ENVIRONMENTAL IMPROVEMENT PROGRAMS

Roosevelt's environmental legacy also included perhaps the most ambitious agenda of environmental "improvements" of any presidency in American history. From water-management structures to roads, trails, and buildings in the national parks and national forests, the Blue Ridge and other parkways, and other infrastructures, the New Deal opened up the nation's water resources for multipurpose management and its natural heritage for the American people to experience and appreciate. It also provided the facilities necessary for human access and recreational use of the national parks, forests, and recreation areas, and thus laid the foundations for the mass public support for preservation of these areas that has emerged in the modern environmental era.

Most far-reaching among these were the New Deal's large multipurpose water management projects, particularly the Tennessee Valley Authority (TVA) projects and also others such as Boulder Dam on the Colorado River, Grand Coulee and Bonneville on the Columbia, and Fort Peck Dam on the Missouri. Roosevelt also championed the construction of the St. Lawrence Seaway, and (unsuccessfully) a Missouri Valley Authority to be modeled on the TVA (Leuchtenburg 1995, 188–95). On a smaller scale but more ubiquitously, the SCS subsidized smaller dams on tributary streams as well, for agricultural and community water management.

The specific motives for these water-management initiatives combined multiple agendas: Progressive desires to use public hydropower production as an alternative "yardstick" to private utility pricing, and additional New Deal aspirations of providing cheap power to electrify rural communities and farms, providing public jobs and purchasing power to the unemployed, "priming the pump" of the economy more generally, and others. The unifying vision used to promote them, however, was an ideal of harmonious human management of the natural environment that has since been largely forgotten or rejected in the United States. In Roosevelt's words, speaking of the Muscle Shoals facility that was to become the nucleus of the TVA:

> There we have an opportunity of setting an example of planning: tying in industry and agriculture and forestry and flood prevention, tying them all into a unified whole over a distance of a thousand miles so that we can afford better

opportunities and better places for living for millions yet unborn in the days to come. (Montgomery, AL, January 21, 1933, *Public Papers* I:888)

Multipurpose river basin management was thus idealized as an extraordinary opportunity to convert natural hazards into productive assets, and to develop productive farms and human communities harmoniously within their natural environment. Instead of devastating floods downstream, dams would capture and store floodwaters for later release for beneficial human uses: clean and cheap hydroelectric power, fertilizer production, urban water supply and irrigation to protect against droughts, and low-flow augmentation during dry seasons to maintain navigation and dilute concentrated pollutants. While the water was stored, the reservoirs would also expand recreational opportunities for swimming and boating, more productive fisheries, and attractive waterfront sites for communities.[2] And in the process, the financial subsidies invested in construction would increase local purchasing power for reconstruction of the regional economy and for longer-term prosperity (Schlesinger 1965, 319–34; Nixon 1957, I:438–40).

By the time of the modern environmental movement in the 1970s, both the substance and the administrative decision procedures of these programs had become the targets of intense attack by environmental advocacy groups (cf. Tarlock, chapter 6). Dam building and stream channelization became two of the great symbolic targets of public opposition. In the name of saving nature from human despoliation by self-interested bureaucracies, the Army Corps of Engineers, Bureau of Reclamation, and even the SCS run amok. In many cases such opposition was well founded, since by then the most favorable projects had been built, the economic depression and soil erosion that justified such projects in the 1930s were long gone, and many of the remaining proposals were more for pork-barrel political purposes than for broader public benefits. Many other valuable functions of natural streams had since been identified as well: floodplains and oxbows slowed and absorbed floods themselves, depositing valuable silts from upstream for agricultural use, and natural stream fisheries were more valued by many fishermen than the species common to man-made lakes. Over time, dams would also silt up; what then? But amid the droughts and floods of Roosevelt's 1930s, these river-basin management projects represented a far more positive ideal for harmonious management of human interactions with nature than later generations have acknowledged—and in regions such as the Tennessee Valley, they have left a far more positive overall legacy of benefits.

FDR's New Deal public works programs also built countless infrastructure facilities opening the national parks and forests to public recreational use. From parkways, access roads and bridges to trails and beaches, campgrounds and visitor centers and other facilities, a substantial fraction of the facilities available even today in the national parks and forests were built by the public works and federal jobs programs of FDR's New Deal. More generally, the New Deal also left a vast and continuing environmental legacy of other civil infrastructures, from courthouses and other public buildings to airports, bridges, and even wastewater treatment facilities (Schlesinger 1965, 287–88; Owen 1983, 84–85; Leuchtenburg 1995, 256–60).

Many of these programs also were elements of a more overarching New Deal agenda for a "new rural conservation," improving both the amenities of life and more specifically the purchasing power of farm families and rural communities, as Sarah Phillips points out in chapter 5 of this book. Agricultural stabilization and conservation, rural electrification, cheap electricity and abundant water, restoration of eroded and deforested lands—all were elements of a broad New Deal agenda for improving the economic and environmental conditions of rural life, and restoring a larger share of purchasing power and economic benefits to farming families and rural residents.

The characterization of such initiatives as environmental "improvements" is no longer common nor consensual: modern environmentalism is far more skeptical of human claims of "improving" nature, and far more apt to contrast the beauty of pristine natural areas untouched by human presence with the pollution of industrial production and the ugliness of urban decline and sprawl. In Roosevelt's time, however, infrastructure construction projects combined with reforestation and land restoration evidenced a perspective that more frankly idealized a vision of managed landscapes and river basins, of harmonious integration of human uses and technologies into the natural landscape. They also combined this vision with the priorities of environmental restoration, economic recovery, and jobs and self-respect for individuals who needed them. As Roosevelt himself noted, in a message to Congress in 1935:

> It is an error to say we have "conquered Nature." We must, rather, start to shape our lives in a more harmonious relationship with Nature. (January 24, 1935, in a message sending to Congress the first report of the National Planning Board, *Public Papers* IV:62)[3]

Such a statement could as easily have been used to introduce the National Environmental Policy Act of 1969, that great legislative policy statement of the modern environmental era, or the 1987 report of the United Nations' World Commission on Environment and Development, which articulated "sustainable development" as the essential goal for governments worldwide.

INSTITUTIONAL INNOVATIONS AND INCENTIVES

FDR's legacy also included an unprecedented range of innovations in environmental management institutions and incentives.

To strengthen federal assistance for soil and water conservation, he transformed the Department of Agriculture's fledgling Soil Erosion Service—a tiny agency, itself created only in 1928—into a new SCS charged to minimize erosion and restore agricultural lands by promoting conservation farming practices. Now renamed the Natural Resource Conservation Service, the SCS provided technical assistance—and beginning in 1954, larger-scale federal funding for tributary dams, stream channelization, and other water management projects—as incentives to farmers to adopt conservation farming and water management practices (Rasmussen, Baker and James 1986, 209–16).

To implement such programs, he relied not on federal regulators but on new models for cooperation and assistance between federal agencies and local user organizations. Examples included the soil and water conservation districts—administered by landowner-elected supervisors—and the rancher-operated local grazing districts. These programs provided federal technical and financial assistance to individuals in exchange for voluntary but verifiable commitments to conservation practices. They also provided new models for organizing and mediating such partnerships through local user organizations. At the same time, however, they also in effect vested privileged use rights and solidified organized influence in these user groups at the expense of other potential stakeholders (Morgan 1965, 37–40).[4]

The agricultural adjustment program was another variant of such innovations. Drafted in consultation with farm leaders, its purpose was to stabilize agricultural prices and production levels in part by encouraging land retirement and conservation practices.[5] Like the grazing districts and other soil conservation programs, the Agricultural Adjustment Act (AAA) was to be implemented through some 5,000 county-level associations of farmers themselves; the government, in Roosevelt's vision, was simply to supply the "unifying element" (*Public Papers* IV:384). Its policy incentives represented a radical innovation, promoting large-scale agricultural land conservation as a method for reducing overproduction and promoting a healthier agricultural economy while also protecting and rebuilding the land itself.

The environmental legacy of the agricultural adjustment programs is mixed (Paarlberg 1989; Rasmussen et al. 1976). In practice, while they encouraged land retirement and other conservation incentives for some lands, over the long term they also promoted overproduction and intensified use of mechanization and chemicals on acreage that remained in production, since they guaranteed government purchase of surpluses to maintain parity prices. They also promoted monocropping, consolidation of farming into large agribusinesses, and vast expansion of federal subsidies in pursuit of votes in the farm states. In many cases these forces even led to destruction of some of the original conservation practices in favor of more intensive production methods (center-pivot irrigation in place of shelterbelts, e.g., causing groundwater depletion in many areas). Finally, they produced a continuing legacy of crop surpluses to be marketed overseas, thus not only benefiting the balance of trade and urban consumers elsewhere but also depressing crop prices for indigenous farmers in poor countries. Roosevelt probably neither intended nor foresaw many of these long-term effects, but the program remains an important element of his legacy.

Finally, Roosevelt's legacy included major though incomplete steps toward creating more effective institutions for Executive policy coordination and planning. Early on, as the major dam-building initiatives got underway, the Fish and Wildlife Coordination Act of 1934 mandated formal interagency consultation between the water and wildlife agencies on the impacts of water projects on fisheries, and on measures such as fish ladders to mitigate them. The result was a first step toward more explicit coordination of planning across the growing number of otherwise fragmented, mission-oriented agencies.[6]

Roosevelt's Committee on Administrative Management (the "Brownlow Committee") made more far-reaching recommendations for administrative reform and reorganization in its 1937 report, including reconfiguring the Department of Interior as a Department of Conservation.[7] This latter proposal was blocked by agricultural and forestry interests and ultimately defeated, however.[8] Roosevelt did succeed in creating the U.S. Fish and Wildlife Service by presidential reorganization, merging a competent but weak Bureau of Fisheries in the Interior Department with the Agriculture Department's more problematic Biological Survey, which had become controversial for its preoccupation with exterminating "pest" species for farmers and ranchers. Roosevelt's merger of these—under the leadership of the respected conservationist Jay "Ding" Darling, who had already begun to reform the Biological Survey—created a respected lead agency for management of wildlife and their habitats.

The New Deal wildlife management legacy also included key innovations in funding that had been championed by Darling, the Duck Stamp Act of 1934 and the Pittman–Robertson Act of 1937.[9] These laws provided dedicated sources of funding, supported by user fees on hunters and fishermen, for wildlife research and management. Wildlife conservation was a perennially weak priority for federal budget support, and the creation of a dedicated, user-supported funding source thus provided both a more stable source of revenues and a more active political constituency for it than had previously existed. Like the grazing fees charged to support range restoration and management, however, the creation of such relationships also tended to privilege the priorities of fee-paying constituencies—hunters and fishermen—over other stakeholder groups that might prefer funding for endangered rather than sport-hunting species, for other species' habitat management, or for other priorities entirely.[10]

Roosevelt also transferred jurisdiction over historical and military sites to the National Park Service, and gave it jurisdiction over many national recreation areas around the new lakes being created by federal water-management projects. He thus consolidated the Park Service's lead role in managing the preservation of natural and cultural heritage sites as well as public access to many federal recreation and tourist areas.

Finally, at the Executive level Roosevelt strove intently but ultimately unsuccessfully to create permanent executive-level planning and oversight institutions. Examples included a National Planning Board and several successors: National Resources Board, National Resources Committee, and ultimately a National Resources Planning Board housed in the Executive Office of the president. Roosevelt's intent in creating these organizations was to strengthen the government's capacity and the president's leadership for crosscutting initiatives to solve complex problems at all geographic scales. Each in turn however faced entrenched resistance from mission-oriented agencies and their client constituencies, and from congressional opponents of strong presidential management more generally, and each ultimately was undermined by congressional refusal to fund them. Roosevelt did succeed in creating the modern Executive Office of the President, including the fiscal planning capabilities of the Bureau of the Budget (relocated from the Treasury Department), but he was otherwise unsuccessful in instituting a permanent presidential planning capability across the fragmented and client-based agencies of the Executive branch (Graham 1976, 49–68).

IDEAS, PRINCIPLES, AND VALUES

Beyond these specific changes in environmental programs, incentives, and management institutions, Roosevelt left a broader legacy of ideals and principles. FDR has often been characterized as far more a pragmatist and experimenter than an advocate of a preconceived ideology. However, his New Deal environmental initiatives exemplified distinctive and consistent overarching principles.

One principle was a commitment to activist presidential leadership and government initiatives to address immediate and pressing problems, both economic and environmental. In the environmental area as in others, Roosevelt used the powers of government not just as a limited and apologetic intervention in the market, but as an active, essential, and legitimate complement to market forces to achieve a good society. Keynesian economic "pump-priming" was one element of this position, but so were other explicit policy goals: revitalizing prosperity in rural communities and farm households, restoring damaged land and forest resources, efficient multipurpose use of water resources, providing jobs for the unemployed and social safety nets for the poor, and promoting price competition in the electric utility sector. Like the Progressives, though unlike many recent presidents, FDR believed in the use of government as a positive force to achieve a broad public interest that included markets but was not limited to market behavior alone. Beyond the Progressives of earlier periods, he also believed in the use of government powers not simply as a counterweight to corporate power, but to stimulate economic and environmental recovery from collapse, and to stabilize both the economy and Nature—and vulnerable individuals and groups in particular—against the destructive extremes of market forces. Perhaps nowhere was this principle better exemplified than in FDR's energetic leadership in multipurpose water resource management in the public interest, and organized as a public utility.

A second core principle of FDR's legacy was a willingness to experiment with new policy innovations to solve environmental problems. These included federal subsidies for environmental restoration and conservation, and for construction of multipurpose water management projects and infrastructure facilities; user fees to support fish, wildlife, and range management; and widespread promotion of partnerships with local environmental user organizations, exchanging federal services and financial assistance for voluntary adoption of conservation practices and for expanded user-group roles in both implementation and future policy influence. In contrast to the modern "environmental era," and even to Roosevelt's own initiatives in other economic sectors (such as industry, labor, banking and investment), New Deal environmental policies included almost no reliance on federal regulatory mandates. Rather, they relied on mobilizing cooperative initiatives between the federal government and local citizens and organizations, and provided a framework of incentives for voluntary adoption and participation, administered and enforced by local user associations. While such subsidies and partnerships had some negative consequences, such as the entrenchment of beneficiary constituencies at the expense of other stakeholders, they were in many respects effective tools for their intended purposes as well.

A third defining principle was Roosevelt's personal vision of integrating economic and environmental restoration, and in particular his personal commitment to the importance of improving rural living conditions by promoting both land conservation practices and active natural resource management. No other president has focused so personally or so effectively on the restoration of environmental damage, nor on the principle that such restoration is integral and beneficial to economic recovery rather than in conflict with economic priorities.

A fourth key principle for Roosevelt was his deep and enthusiastic belief in the value of planning. As noted in the quotation above about Muscle Shoals, Roosevelt believed passionately in regional-scale integrated planning to create what we would today call sustainable development: communities at the scale of large watershed regions that were economically and socially as well as environmentally healthy, tying together agriculture and forestry, industry and human communities, water resources and human needs. His vision of such planning was not rigid central control, but required strong advisory institutions at the Executive level—the National Resources Planning Board (NRPB), for example—to provide the crosscutting perspective and integrated information necessary to conceive integrated solutions. NRPB's authority was simply to conduct studies, produce reports, and propose recommendations: the power simply to gather and present information on important crosscutting problems, and to place them on the national agenda. No other modern presidency has created such a capacity, with the modest exception of the Council on Environmental Quality—created by legislative rather than presidential initiative—during its initial decade.[11]

Finally, a fifth central principle of FDR's environmental legacy was his commitment to improving the effectiveness of government administration itself. As he put it, in a speech on his philosophy of governance in September 1932:

> The day of the great promoter or the financial Titan . . . is over. Our task now is not discovery or exploitation of more natural resources, or necessarily producing more goods. It is the soberer, less dramatic business of administering resources and plants already on hand, of seeking to reestablish foreign markets for our surplus production, of meeting the problems of underconsumption, of adjusting production to consumption, of distributing wealth and products more equitably, of adapting existing economic organizations to the service of the people. The day of enlightened administration has come. (San Francisco, CA, September 23, 1932, *Public Papers* I:752)

For all his reputation as an improviser, in short, a major element of FDR's environmental legacy was the creation of important new administrative agencies for environmental management tasks, the reorganization, expansion, and modernization of others, and the attempt to create an effective Executive structure for coordinating and integrating across them.

ENVIRONMENTAL RESULTS

The immediate results of Roosevelt's initiatives for the environment were overwhelmingly positive. For soil conservation, reforestation, wildlife management, and many other environmental restoration programs there would be no serious disagreement on this. For multipurpose water projects and other infrastructure construction, many environmentalists of the modern era might well argue that all free-flowing rivers and natural areas would have been better left pristine; but one could at least as persuasively argue that for most of the projects actually built during the New Deal, their environmental as well as economic benefits outweighed what was given up. Hydroelectric power was a far cleaner alternative to fossil fuels, planned and integrated river basin management did help to create more harmonious and sustainable patterns of development than the ad hoc development and abandonment patterns that preceded it, and increased public access and facilities in the national parks and forests—not to mention the personal experiences of CCC and public-works workers and others—helped immeasurably to create the mass constituency of support that made possible the modern environmental movement.

Three other elements of FDR's legacy had more problematic consequences, however, and it is these rather than the direct environmental effects of the New Deal programs that have more legitimately deserved criticism by modern environmentalists.

The first of these was the consequences not of FDR's discretionary initiatives but of the exigencies of World War II. This was also FDR's legacy, though driven more by necessity than by voluntary choice. The imperatives of the war required all-out industrial mobilization, including unprecedented increases in levels of materials and energy extraction, industrial production, and relocation and concentration of the population from rural areas to jobs in urban and industrial regions. Once this economic infrastructure had been created, however, it would not simply fade away after the war, but would demand new markets to maintain both itself and the jobs of those now dependent on it; and the Cold War arms race would be used to continue to justify maintaining its capacity for military purposes as well. The imperatives of war production—the construction of massive industrial capacity for manufacturing transport vehicles and aircraft, durable goods, synthetic chemicals, munitions, and other war needs, the energy supplies to power them, and the food demands of both American troops and war-torn allied nations—thus created a permanent shift in the nation's economy toward greatly increased extraction and pollution pressures on the environment.

It is also only fair to note, however, that this wartime production capacity and its postwar momentum, coupled with associated increases in individual incomes, also helped to create a mass middle class affluent enough not only to buy suburban houses, cars and appliances, but also to care about environmental problems, and to mobilize political demands for pollution control, endangered species protection, wilderness preservation, and other environmental values that were previously and elsewhere given low priority.

The second problematic legacy was the postwar excesses of some of the initiatives set in motion during the New Deal period, as they took on continuing political momentum of their own. The integrated river basin management model of FDR's New Deal, for instance, was replaced after the war by a billion-dollar-a-year nationwide program of dam building; and the SCS's programs shifted in emphasis from land conservation measures to dam and stream channelization subsidies. Together, these projects ultimately dammed nearly every major reach of river in the country, and destroyed many others by stream channelization.[12] The agricultural stabilization program evolved from a program aimed at supporting family farms and rural communities into a massive and continuing subsidy program for a small number of staple-crop agribusinesses. And national forest management practices evolved during and after the war into large-scale industrial clear-cutting, benefiting the commercial logging companies but at the expense of other values and uses of the national forests.

It is common but perhaps unfair to blame these later excesses on Roosevelt and the New Deal, particularly since they were so contrary in their effects to both the environmental goals and the environmental achievements of the New Deal itself. In reality they were driven not by New Deal ideals or presidential leadership but by congressional pork-barrel politics.

A third and more ambiguous legacy of the New Deal, however, was the political legacy of the New Deal's promotion of partnership programs with organized client groups such as farmers, ranchers, and others. These partnerships (and their associated federal subsidies) were quite effective in promoting conservation practices in the short run, and thus in restoring and conserving land—certainly far more widely and effectively than would have been accomplished by any federal regulatory program. They also broadened access to federal programs and decisions, beyond the major corporations and financial institutions that already had privileged access and influence: farmers, ranchers, and some other stakeholder constituencies now had more organized access and influence as well (Leuchtenburg 1963, 89). At the same time, however, these partnerships in effect privileged (and in some cases, created) a few additional vested interests, at the expense of the overall public interest that both Progressives and New Dealers advocated in principle.

The postwar residue of the New Deal environmental programs, therefore, was the individual agencies without the planning or policy integration, fragmented into narrow bureaucratic "subgovernments" and "iron triangles" serving organized interests at the expense of less organized constituencies.[13] In practice, the "iron triangle" alliances among agencies, clients, and congressional committees became profoundly resistant both to other legitimate interests and to changing public values and priorities. It was this gap between rhetoric and reality—between the ideal of disinterested expert planning in the name of an overarching public interest, and the perception of a fragmented and client-dominated "broker state"—that motivated the environmental movement to attack the Progressive governance model itself, and in its place to advocate expansion of citizen rights to demand full access to federal records and to challenge federal actions in the courts (Andrews 1999, 218–21).

Finally, some New Deal initiatives had longer-term consequences that were not clearly foreseen at the time. The introduction of the federally insured, self-amortizing mortgage made the American dream of a single-family suburban house possible for millions of families, but also promoted both suburban sprawl and inner-city decay (Jackson 1985). Agricultural adjustment subsidies in the long run increased rather than reduced surpluses, growing them on less acreage but with more intensive impacts of pesticides, fertilizers, and mechanization—and depressed farm prices in developing countries as well. Federal subsidies for hydroelectric power and other energy sources encouraged an economy that was structurally dependent on cheap energy, and more vulnerable to supply disruptions and unexpected price fluctuations. Even the TVA evolved from its visionary model of harmonious river-basin development into simply a powerful and politically insulated electric utility, committed far more to strip-mined coal and nuclear technology than to clean hydropower.

FDR's LEGACY IN THE MODERN
"ENVIRONMENTAL ERA"

The coalescence in 1970 of a nationwide "environmental movement" was one of the most dramatic examples of mass political mobilization of twentieth-century American politics, and created the fleeting possibility of a modern version of a New Deal for the environment. The National Environmental Policy Act (NEPA) passed by both houses of Congress in 1969 with less than 20 dissenting votes, and was signed by the president on television on New Year's Day 1970 with a statement hailing it as the start of a "decade of the environment," in which the time was "literally now or never" to clean up the pollution of decades of industrialization and urban growth.

NEPA was the most clearly New Deal-like of the modern environmental statutes: it combined a policy statement, "action-forcing provisions" and a presidential-level Council on Environmental Quality, aimed squarely at the unfinished New Deal goals of harmonizing human actions with their consequences for nature, of coordinating and integrating the actions of the diverse federal agencies to that end, and of providing an executive-level staff institution—a modern National Resources Planning Board, of a sort—to advise the president, monitor the agencies, and conduct agenda-setting studies on crosscutting issues.[14]

Four months later, the original Earth Day produced an unprecedented outpouring of grassroots public support for government action to protect and restore the environment, the largest public demonstrations since the victory parties at the end of World War II. Nationwide and nonpartisan, it created the political conditions to support enactment of an unprecedented series of tough federal environmental regulatory statutes, as well as multibillion-dollar subsidies for wastewater treatment facilities and cleanup of contaminated sites. These laws were passed by broad bipartisan majorities over the decade of the 1970s, and signed by both Republican and Democratic presidents. As Tarlock (chapter 6) and others have pointed out, they also incorporated mechanisms for universalizing citizen access to government information and to the courts, so that not just vested interests but any interested individual or group could challenge either industry or government agencies themselves over pollutant emissions, other environmental impacts, or failure to implement or comply with the environmental statutes. These mechanisms reflected the post–New Deal cynicism toward agency capture by vested interests, and provided new safeguards against it; but with this new twist, they provided an extraordinarily broad new base for presidential environmental leadership, comparable only to the broad base of support FDR himself had available amid the economic and environmental disasters of the 1930s.

Comparable leadership did not materialize, however. Even Jimmy Carter ran against government and rejected the New Deal model, rather than using this opportunity for presidential leadership (Leuchtenburg 2001, 191–95), and Ronald Reagan set out not to reform or integrate federal environmental policy-making but to fundamentally dismantle it.[15] Since Ronald Reagan's election in 1980, a conservative counterattack has reduced the environmental policy agenda to ideological trench warfare and stalemate. Reagan's attack on the environmental regulatory programs not only galvanized Democratic electoral gains in 1984, but also redefined the environmental movement as predominantly a partisan constituency for the Democrats. By cutting off federal subsidies for wastewater treatment plants, he split off local governments from the environmental coalition, leaving them with unfunded mandates rather than with a balance of benefits and burdens.[16] While Reagan seriously overreached in attempting to dismantle the environmental regulatory programs, overall the resurgent right has largely succeeded in promoting public cynicism toward government, in redefining the environmental agenda from a positive shared vision for society to a negative image of onerous and meddlesome government regulation, and in at least stalemating, though not dismantling, the federal environmental policy regime.[17]

Aside from NEPA, moreover, the environmental agenda of the 1970s was different in key respects from that of the 1930s. For both, a central priority was environmental cleanup and restoration: from soil erosion, deforestation, and floods in the 1930s, and from air and water pollution and toxic hazards in the 1970s. The New Deal, however, pursued this using subsidies, jobs, partnerships, and infrastructure investments rather than regulation as its primary tools: it regulated businesses' economic practices, but sought most fundamentally to revive and reenergize them. The environmental statutes of the 1970s, in contrast, were more fundamentally regulatory, rooted in popular distrust and moral disapproval of corporate polluters rather than in economic incentives to assist them. The New Deal built popular support by providing direct benefits and protections to ordinary people, often for the first time (Leuchtenburg 1963, 326 ff.); the environmental initiatives of the 1970s benefited them only more indirectly, by regulating corporate polluters, and eventually by actions regulating their own behavior (driving and car inspections, for instance) and that of their local governments and small businesses (trace contaminants in drinking water, underground storage tanks, wetlands and endangered species habitats, and others). Roosevelt and his brain trust led many of the New Deal legislative initiatives for conservation; in the 1970s most initiatives originated in the Congress and in environmental interest groups, while successive presidents acquiesced in their enactment and their agencies were largely preoccupied with writing the many regulations and thousands of permits they required.

As the initial attention of the mass public inevitably subsided, the environmental movement itself gradually became trapped in a predominantly oppositional role rather than building broader alliances and new initiatives toward a common positive vision comparable to that of the New Deal.[18] It differentiated into a network of mainstream interest groups supporting expert staffs and using both lobbying and litigation as tools of strategic policy influence, ad hoc grassroots groups fighting more local issues (environmental justice groups, for instance), and a more passive general public that could still be mobilized through the media to object to specific outrages (new environmental crises, as well as proposals to roll back existing environmental statutes or regulations) but otherwise were mainly checkbook or passive supporters. The core of the movement also became more active in electoral campaigns, predominantly on the Democratic side, and many Republicans in turn wrote off any likelihood of their support and turned to more conservative groups as their base of activists.[19]

The environmental groups' challenge ever since has been to keep proving their ability to mobilize a larger mass public—at a level that matters in elections—in order to maintain their influence, in the face of conservative and business propaganda painting them as elitists, as advocates of big-government regulation and taxation, and as naïve extremists. Distrustful of both businesses and government, and committed to adversarial methods such as regulation, litigation, and public campaigns against symbolic villains, they have found this an increasingly difficult and defensive challenge except when events have provided new crises and media attention to reenergize mass public support.[20] They also have largely failed to build strong long-term alliances with other constituencies, such as organized labor, minorities, and advocates for the poor.[21]

Perhaps the deepest challenge to the modern environmental movement is its lack of a positive vision equivalent in both coherence and popular support to that of the New Deal, or of previous environmental movements such as Theodore Roosevelt's Progressive conservation, the "City Beautiful" movement of the 1890s, or the urban sanitation movement of the mid-nineteenth century.[22] The closest modern equivalent is arguably the vision of "sustainable development" articulated by the 1987 report of the World Commission on Environment and Development (the Brundtland Commission), which provided the basis for the "Agenda 21" of the United Nations endorsed at the Rio "Earth Summit" in 1992 (WCED 1987). This report and Agenda 21 itself articulated in considerable detail a vision of future development that combined economic progress with both ecological sustainability and social equity. Despite some issues requiring further debate, it comes the closest to any modern document of stating a vision of a New Deal for the world. Within the United States and even among U.S. environmental groups, however, it has been largely ignored, and has had little impact on agendas or priorities except in a very few local governments.

WHAT WOULD FRANKLIN DO?

Are there lessons from the legacy of FDR and the New Deal that still have merit today? Or are we now in an era to which they are no longer relevant, or in which they have been discredited by their own history in favor of more conservative and "market-based" solutions? Putting it another way, what would FDR do today, and would his solutions be more effective than those now in use?

In domestic environmental affairs, Roosevelt could undoubtedly have given vastly more effective presidential leadership to the environmental agenda in the 1970s, when the major environmental statutes were being introduced and the environmental agencies reorganized. Armed with NEPA, with the other major environmental statutes, and with the vast reservoir of mass public support that was then available to shape the national agenda and its implementation, Roosevelt could have been far more effective in consolidating the gains of the "environmental era," integrating (and in some cases, reform) the environmental regulatory programs, and thus defusing much of the antiregulatory backlash that Reagan used so successfully to redefine the agenda.[23] He clearly would have used the authority of NEPA to integrate environmental with economic and social policy-making with greater vision at the Executive level, and to articulate a positive vision for integrating them rather than relying on regulation alone.

Roosevelt might also have managed the energy crises of 1973 and 1979 very differently, perhaps with bold initiatives to promote the development and widespread adoption of energy conservation and alternative fuels, to reenergize public optimism and stimulate the economy out of the "stagflation" of that era, and to coordinate the fragmented institutions of energy policy and integrate them more effectively with those of other agencies and sectors. One could perhaps even imagine integrated energy planning and policy incentives today as a modern counterpart to his rural electrification and river-basin planning initiatives. How he would have dealt with the later consequences of his own farm subsidies is more difficult to speculate. And whether he could have been as effective today, in the wake of the resurgent conservative Republicanism of the 1980s and 1990s, would depend in part on whether events were again to provide opportunities to mobilize widespread public support.

Internationally, however, there is probably little doubt about what sorts of environmental policies Roosevelt might have pursued. FDR was clearly an internationalist, who lived to see the founding of the United Nations and championed the Bretton Woods agreements creating the International Monetary Fund and the World Bank. He clearly viewed these institutions as sources of more jobs and a higher standard of living through trade, and more broadly as foundations of a New Deal for the people of the world: as tools for reconstruction rather than merely relief, to promote worldwide recovery from war by putting purchasing power in the hands of ordinary people as well as managing more systemic factors—such as exchange rates—that only broader institutions could address (*Public Papers* XIII:548).

These institutions too have had controversial and problematic consequences, both environmental and economic, that were not foreseen by Roosevelt and their other architects. In economic terms, the income gap between the richest 20 countries and the poorest 20 has doubled since 1970: the average annual income in the richest countries is now 37 times higher than in the poorest, mainly because of lack of growth in the poorest countries. The poorest countries went from a growth rate of about 2 percent annually in the period 1960–1980, to a *decline* of 0.5 percent per year between 1980 and 2000; and more than three-quarters of all countries saw their per capita growth rates decline by at least 5 percent from the period 1960–1980 to 1980–2000. The absolute number of people still living in extreme poverty—at less than $1 a day—includes nearly a quarter of the population of developing countries. While poverty rates have improved in East Asia and especially China, they have worsened in Latin America and sub-Saharan Africa, and have worsened dramatically in Russia and Eastern Europe (World Bank 2003).

Across large areas of the world, environmental conditions also are deteriorating. One-third of the world's people live in countries that are experiencing moderate to high water shortages, and that proportion could rise to more than half in the next 30 years. More than a billion people in low- and middle-income countries lacked access to safe water for drinking and personal hygiene in 1995, and hundreds of cities in developing countries have unhealthy levels of air pollution. More than 20 percent of all cropland, pasture, forest, and woodland have been degraded since the 1950s, 16 percent so severely that the change is too costly to reverse. An estimated 58 percent of the world's coral reefs and 34 percent of all fish species are at risk from human activities, and fully 70 percent of the world's commercial fisheries are now fully exploited and experiencing declining yields.

In short, while the world economy in aggregate has created unprecedented wealth and amenities for a few, it is failing to meet even the basic needs of a large fraction of the human population or to protect its natural resources and the ecosystems that produce them.

It seems likely that Roosevelt today would seize on the sustainable development agenda as a contemporary vision of a New Deal for the world, and as an urgent priority, would offer far stronger U.S. leadership for integrated economic, environmental, and social reconstruction in partnership with other nations and with the peoples of the world.

CONCLUSION: LEGACIES AND LEADERSHIP

Tarlock (chapter 6) and others argue correctly that modern environmentalists often see FDR's legacy more as the source of many modern problems than of solutions, and tend to reject both the hubris of trying to manage and "improve" nature for human betterment and the procedures of rational planning by an expert-run administrative state. It is also true that modern environmentalists have derived much of their greatest power and influence in environmental policy from their use of the rights of challenge to the federal agencies using the adversarial procedures provided by the Administrative Procedures Act, the law passed in 1946 by postwar conservatives to oppose and constrain the New Deal administrative agencies.

At the same time, the legacy of the New Deal itself offers far more significant positive accomplishments for the environment than any other presidency before or since, and many (though not all) of the problems often associated with it spring more from later excesses than from its immediate environmental outcomes. Modern environmentalists themselves are now under severe challenge as well, and face the constant and increasing difficulty of maintaining their influence as mass public support waxes and wanes amid other larger issues, such as terrorism and homeland security, war, and economic recession. They also face the reality that resurgent, well-funded, and effectively mobilized conservative interest groups have stalemated their positive agenda and forced them increasingly onto the defensive simply to protect past gains. The policy achievements of the modern environmental era, like many of those of the New Deal as well, have not been dismantled, but they are under constant and increasingly sophisticated siege.

It is timely, therefore, to consider what positive lessons modern environmentalists might learn from both the policies and the politics of Roosevelt's legacy. Certainly one would not advocate any mere re-creation of an earlier and different era, and there is no obvious candidate for presidential leadership in sight comparable to FDR in any event. But both FDR's policies and programs, his vision and his innovations in accomplishing it, bear serious consideration for the present and future.

NOTES

1. Many of the details recounted here were previously published and more fully documented in Andrews 1999, chapter 9.
2. The New Deal explicitly developed and told this story in several influential films, such as *The Plow That Broke the Plains* and *The River,* which interpreted the environmental disasters of the Dust Bowl and floods and the New Deal's responses to the American people. Publicity for the TVA and other initiatives also show these themes. The millions of young men who worked in CCC camps and on other New Deal public works projects, of course, absorbed these ideals through their personal experiences as well, as did thousands of farmers who received New Deal assistance to introduce conservation farming and reforestation practices.
3. In this message Roosevelt also notes that he uses the term *national* resources to include not only natural resources such as land and water, but also people interacting with Nature in the course of their occupations and urbanization.
4. TVA also developed extensive use of partnerships with local farm and community stakeholder organizations.
5. Agricultural Adjustment Acts of 1933 and 1938, Soil Conservation and Domestic Allotment Act of 1936.
6. The National Environmental Policy Act of 1969, with its requirements for environmental impact statements and for comprehensive interagency review of them, was originally introduced in the House as an amendment to the Fish and Wildlife Coordination Act.
7. The committee recommended consolidating some 100 departments, boards, commissions, and other separate units into 12 administrative departments, including renaming Department of Interior as a Department of Conservation and adding new departments of Social Welfare and Public Works. It also recommended strengthening the staff and analytical capacity available to the president, putting the entire administrative service on a merit basis, and creating an auditor general to make the Executive branch more accountable to Congress.

8. The transformation of Interior into a Department of Conservation has never occurred, despite a later proposal by President Richard Nixon's Ash Commission as well. As a second-best alternative, Nixon created the Environmental Protection Agency (EPA) in 1970 by a presidential reorganization plan that consolidated only pollution-related regulatory functions.

9. Later augmented by the Dingell–Johnson Act of 1950.

10. Similarly dedicated funding sources are often opposed by environmentalists in other cases, such as the Highway Trust Fund that dedicates gasoline tax revenues primarily to the support of road construction rather than mass transit and other transportation alternatives.

11. In the 1970s, the President's Council on Environmental Quality (CEQ) (created by the National Environmental Policy Act (NEPA) of 1969) produced a number of influential agenda-setting reports on topics such as toxic substances, off-road vehicles, agricultural lands, and global environmental problems, among others. President Reagan however reduced CEQ's budget and staff in 1981 to the point that this function was no longer possible, and replaced its entire staff with persons who had no interest in continuing it.

12. Even with the unemployment and flood crises of the Great Depression long gone, e.g., and the economy in high gear, Congress in the 1950s subsidized thousands of dams and other water-management projects at the behest of individual states and legislators. They refused however to create further basin-wide planning and management agencies modeled on TVA, such as proposals for the Missouri and Columbia Rivers.

13. The empirical political science literature of the postwar decades produced a rising chorus of critiques of interest-group-dominated federal environmental agencies. See Andrews 1999 218 n. 27.

14. NEPA's policy statement, for instance, reads: "The Congress, recognizing the profound impact of man's activity on the interrelations of all components of the natural environment, and particularly the profound influences of population growth, high-density urbanization, industrial expansion, resource exploitation, and new and expanding technological advances . . . declares that it is the continuing policy of the Federal Government . . . to use all practicable means and measures . . . in a manner calculated to foster and promote the general welfare, to create and maintain conditions under which man and nature can exist in productive harmony, and fulfill the social, economic, and other requirements of present and future generations of Americans."

15. His policy for EPA, e.g., was to "deregulate, de-fund and devolve" its programs to the maximum extent possible through unilateral Executive actions; and he did away with the Council on Environmental Quality in all but name, firing its entire staff (even the secretaries) and replacing them with a skeleton crew to continue its minimum statutory functions.

16. The second round of environmental regulatory statutes passed by the Democratic majority after 1984 also imposed far greater mandates on local governments and small businesses than did the original statutes, most of which were aimed more at large corporations. These statutes were motivated by growing recognition that many pollution problems originated with these

smaller but more widespread sources, but the political risks probably were not as well recognized.

17. The only major environmental legislation to pass the Congress since the 1980s, e.g., was the Clean Air Act of 1990—though some might add the Food Quality Protection Act of 1996—and the recommendations of multiple bipartisan blue-ribbon commissions for constructive reforms have all failed, due to the pervasive distrust between pro- and antienvironmental factions.

18. In this it is perhaps more similar to the issue politics of Theodore Roosevelt's era, mobilized by the muckraking journalists' moralistic attacks on corporate abuses, than by the more distributive politics of the New Deal conservation programs.

19. Exceptions include a few remaining Republican moderates from states with strong environmental constituencies.

20. Environmental groups and EPA have often sought to build broader issue-specific alliances with those businesses and sectors that see their own interests as allied with the environmental agenda—on global warming, e.g.—but these remain largely ad hoc and wary. Some environmental groups that have participated in partnerships or stakeholder negotiations with businesses have been roundly denounced by others for consorting with the enemy.

21. Minorities and low-income communities are often among those most impacted by exposure to environmental contamination and other hazards, but their "environmental justice" advocacy groups have remained largely separate from and less influential than the mainstream environmental interest groups. The environmental groups began to weave an alliance with labor in the 1990s, in opposition to the perceived effects of trade liberalization on both labor and the environment, but this initiative dissipated with the shift in national attention and priorities in the wake of the wars on terrorism and in Iraq.

22. Leuchtenburg, e.g., argues that for all its image of anti-utopian pragmatism in experimenting with ad hoc policy innovations, the New Deal did also have a vision of the "heavenly city" as inspiration, in the image of attractive greenbelt towns nestled into reforested and well-managed regional agricultural landscapes and small, clean factories. TVA's "model village" in Norris, Tennessee also provides an example (Leuchtenburg 1963, 345).

23. There was in fact a credible bipartisan consensus developing in the 1970s around ideas for reform and integration of environmental regulations; Reagan chose instead to attempt to dismantle them.

REFERENCES

Andrews, Richard N.L. 1999. *Managing the Environment, Managing Ourselves: A History of American Environmental Policy.* New Haven, Connecticut: Yale University Press.

Graham, Otis L., Jr. 1976. *Toward a Planned Society: From Roosevelt to Nixon.* New York: Oxford University Press.

Jackson, Kenneth T. 1985. *Crabgrass Frontier: The Suburbanization of the United States.* New York: Oxford University Press.

Leuchtenburg, William E. 1963. *Franklin D. Roosevelt and the New Deal, 1932–1940.* New York: Harper & Row.

Leuchtenburg, William E. 1995. *The FDR Years: On Roosevelt and His Legacy.* New York: Columbia University Press.

Lowitt, Richard. 1984. *The New Deal and the West.* Bloomington: Indiana University Press.

Morgan, Robert J. 1965. *Governing Soil Conservation: Thirty Years of the New Decentralization.* Baltimore: Johns Hopkins University Press.

Nixon, Edgar B., ed. 1957. *Franklin D. Roosevelt and Conservation, 1911–1945.* Washington, DC: U.S. Government Printing Office.

Owen, Anna Lou Riesch 1983. *Conservation Under F.D.R.* New York: Praeger Publishing.

Paarlberg, Don 1989. "Tarnished Gold: Fifty Years of New Deal Farm Programs." Chapter 2 in *The New Deal and Its Legacy: Critique and Reappraisal,* edited by Robert Eden. Westport, Connecticut: Greenwood Press.

Public Papers of Franklin D. Roosevelt, Vols. I–XIII.

Rasmussen, Wayne D. and Gladys L. Baker. 1986. "The New Deal Farm Programs: The Myth and the Reality." In *The Roosevelt New Deal: A Program Assessment Fifty Years Later,* edited by Wilbur D. Cohen. Austin: Lyndon B. Johnson School of Public Affairs, University of Texas, 201–19.

Rasmussen, Wayne D., Gladys L. Baker, and James S. Ward. 1976. *A Short History of Agricultural Adjustment, 1933–75.* Agriculture Information Bulletin No. 391. Washington, DC: U. S. Department of Agriculture, Economic Research Service.

Schlesinger, Arthur M. 1965 [1958]. *The Coming of the New Deal.* Boston: Houghton Mifflin.

World Bank. 2003. *World Development Report 2003.* New York: Oxford University Press.

World Commission on Environment and Development. 1987. *Our Common Future.* New York: Oxford University Press.

CHAPTER 11

A NEW DEAL FOR NATURE—AND NATURE'S PEOPLE

ROGER G. KENNEDY

WE HAVE NOTHING TO FEAR, BUT FEAR ITSELF. Not skepticism, not wariness, but fear. Warily we act prudently. Fear confuses. Terror immobilizes. Terrorists seek to induce terror, and if they cannot achieve immobility in their victims, they seek to induce them to act fearfully, imprudently, and in confusion. That is the lesson of the last 70 years, since Franklin Roosevelt refused to be terrified by the crisis of the 1930s, since Franklin Roosevelt showed how presidents can lead a people away from fear toward a confident, cheerful, and humane society.

The 1930s seem very recent to me; I even recall the voice of Herbert Hoover on the radio when he was president. My political education began in the 1930s, and I was recycled into national politics by President Clinton in the 1990s at the behest of Sidney R. Yates, an old, unrepentant New Dealer, and his masterful assistant, Mary Bain, who had worked for Frances Perkins.

There are numerous scholars who know more about the precise details of the subject at hand than I do. Therefore I do not offer myself to you as the source of any information you do not already possess, but, instead, as an authentic relic of times past, who has had ample time to let his passions simmer and to become unrepentantly opinionated.

I wish to commence with some suggestions arising from this place itself, where Franklin gave as his profession—"forester." He was a great dissembler when the occasion warranted denying to others premature disclosure of his intentions, but in this instance I believe he was describing himself—a little slyly to be sure but quite accurately—as a man who understood forestry— and, therefore, the organic interactions among humans and the other species with which we co-inhabit this earth. This being true, human interaction with nature will be one of the themes of this essay. Not manipulation nor dominance, but interaction—through experiment and openness to feedback. Chaos theory came later—Franklin and Eleanor Roosevelt lived in chaotic circumstances and responded instinctively, not theoretically, by applying to politics the uncertainty principle of physics. They believed in experiment amid complex interactive organic systems—natural and political. Franklin Roosevelt was a *scientific* forester—operating as if he understood chaos theory before it was invented. He loved acting amid chaos—the very complexity of the problems before him, evoked from him what he called "cheerfulness," a state of mind transcending any mood, providing joyous energy to persevere. Anyone who has actually had to conduct public business in a turbulent, changing, and uncertain world, needs a cheerful perseverance to keep on trying, reassessing, repenting, and returning to the fray. Those like the Roosevelts who can act amid uncertainty without souring into meanness or becoming inflamed into ferocity show themselves to be adherents to the creed of kinetic *and hopeful skepticism,* or, if you like, of *steely elasticity.*

Theologically, these terms are grounded in a Neibuhrian and Jungian sense of the imperfection of all humans, including ourselves, and of the improvability of us all, likewise. It is characteristic of democracy because it asserts that no one is wise enough to rule absolutely nor is hardly anyone so depraved as to be deprived of the right to vote. Acknowledging our own insufficiencies, affirming in others a respectable store of competence, we follow the Roosevelts' example when we act in the cheerful, hopeful skepticism that is appropriate in running governments in crisis, in doing heart surgery by candlelight when there has been a power failure, in rescuing mountaineers though the ascent must take place in a fog, in rallying the troops to counterattack before dawn, and in navigating by memory when the charts have washed overboard.

It is, I think, the mood in which we should now draw upon the environmental legacy of the Roosevelts to initiate a New Deal for Nature—and for Nature's People. That proposal, with which I conclude this essay, rests upon the principle that humans are part of nature living as components of its complex interactive system and inescapably bound by its rules. As parts of nature we are also parts of each other. That was the central perception of the New Deal—distinguishing it from regimes holding that persons were merely integers in a market mechanism. The New Deal put people first—not property. Property rests in theory—humanity precedes theory. When a New York real estate developer complained that government-assisted housing for the poor would violate free-market orthodoxy, Franklin Roosevelt responded by quoting Grover Cleveland as saying: "I am faced with a condition and not a theory." The condition was a human condition. People were ill housed and ill fed.[1]

Putting people first and only then property was family doctrine—Theodore Roosevelt set it forth in his speech in John Brown's grove at Osawatomie, Kansas: "The man who wrongly holds that every human right is secondary to his profit must now give way to the advocate of human welfare," he said, and proclaimed that even the judiciary should "be interested primarily in human welfare rather than in property." People first!

This humane imperative was not for Franklin Roosevelt a mandate for arrogance. Though action was necessary, he acted, but never with excessive self-assurance. His skepticality extended to include skepticism about his own exclusive wisdom. When Al Smith turned on him, and attacked him for experimenting, he got an aide to print out for him a quotation from Lincoln to keep on his desk:

> I do the very best I know how—the very best I can; and I mean to keep on doing that until the end. If the end brings me out all right, then what is said against me won't amount to anything. If the end brings me out wrong, 10,000 angels swearing I was right would make no difference.[2]

There it is, Roosevelt's benign and redemptive skepticism—a willingness to act when in doubt, when the welfare of "human beings" demanded action. That phrase comes to mind when we recall how Roosevelt responded to free-market theory, or "policy," on another occasion, when editorialists insisted that it required his cutting off food to the hungry. His answer was that he did not want to hear "the policy" of newspaper owners, but what those writers themselves knew "as human beings"—"as human beings [—] what happens if I cut off relief?" "As human beings" they would understand his experiments, his changes of course, his actions taken with insufficient knowledge. That basis for action was so simple it defied analysis not because it was impenetrable but because it was transparent—it requires no analysis. In the service of human beings, the presidency, he said, is "pre-eminently a place of moral leadership." Uncertain but fearless, he proceeded to act. Each of us of course acts in our public as in our private lives with insufficient knowledge. Franklin Roosevelt was more courageous than most of us because he did not deceive himself about his own insufficiency.[3]

Roosevelt's freedom from self-deception—or, putting it more positively—his candor in his discussions with himself—was the essence of his experimental politics, of his benign, hopeful, cheerful, and kinetic skepticism. I believe that his method of dividing responsibilities among his subordinates may be understood to be his way of coping with the excessive certitudes of others—such as General Johnson, Henry Wallace, and Harold Ickes. Yet he shared with them a commitment to action, in the face of uncertainty. I think that came out of his sense of history—and of the peculiar capacity of humans to think historically. We have memory. We have anticipation. We are, therefore, more responsible than are species without those attributes for what happens to us all—and at the same time inseparable from the other species with which we co-inhabit the earth. Daniel Boorstin helped us understand that point by noting how the Founding Fathers, especially Thomas Jefferson, anticipated the Endangered Species Act: "in his [Jefferson's] writings, we frequently come upon the appropriate verses of the Psalmist, 'O Lord, how manifold are thy works! in wisdom hast thou made them all: the earth is full of thy riches' " And Jefferson himself wrote that "if one link in nature's chain might be lost, another and another might be lost, till this whole system of things should vanish by piece-meal."

Be careful, the psalmist admonishes us. Take account of nature as created—as intentional and as complete. There is no waste in nature. There are no extraneous species, as we are reminded in that famous passage in a sermon of John Donne:

> No man is an island, entire of itself; every man is a piece of the continent, a part of the main . . . any man's death diminishes me, because I am involved in mankind; and therefore never send to know for whom the bell tolls; it tolls for thee.

The tolling of the bell is for the death of any living thing. The voice is that of John Donne, but the echoes come from Saints Patrick and Francis, and from the Buddha. We are "involved" in all life in an intentional and not an accidental universe. In that universe all species and all things, from stars to starfish, have a place.

> . . . if one link in nature's chain might be lost, another and another might be lost, till this whole system of things should vanish by piece-meal.

And so they might, and so they might, one species after another. Unless we rally round each other, and join with all others who acknowledge with us that the bell is tolling constantly now, tolling all day and all night without surcease, as species after species die, creation after creation, friend in the earth after friend in the earth.

* * *

And how shall we of the responsible species—respond to all this? By looking in the mirror before we step outside with our shovels and pickaxes in hand. When we look in the mirror we see there people who are ignorant and so fallible. We are responsible—but we are uncertain—so we will experiment and we will refrain from actions that are likely to have irreversibly pernicious consequences. That is what we mean by steely elasticity. We will not shrink from action, but we will try not to do anything in our half-informed state that cannot be reversed or remedied. We won't be too quick to drain the Everglades, to put a dam across one river or make a concrete sluice of another. We will not arrest the natural purging of mountain slopes by fire. We will not plow soils so thin that they will make a dust bowl, nor clear-cut forested slopes where torrents of rain are likely to leave an eroded wasteland. When we think of the irreversible damage done to this continent, we mostly think of processes set in motion before 1930—leaving aside the terrible errors of continuing to build dams. The second Roosevelt's legacy was mostly a therapeutic one. His agricultural and environmental policies sought to find remedies for past failures and to recognize the principle of avoiding the irreversible. Diffident experiment is the natural method of people doing therapy.

Inescapably all the New Dealers were reminded of the peril of the irrecoverable because they were mopping up the messes left by the heedless overconfidence of an American triumphalism that had transgressed the limits set by nature. They lived as the undeniable evidence of transgression swirled around them in clouds of dust as they traversed eroded landscapes declining into polluted rivers, where once-fertile bottom lands had become uninhabitable industrial wastes inhabited by industrial workers whose very bodies were as ravaged as the hillsides above.

New Deal environmental policy, as carried forward by Harold Ickes, presaged the work of our own time, as we experiment with means to restore stream flow in the Colorado River, restore salmon runs to the Elwha, sort out what to do when rivers fill behind dams with tons and tons of silt, and restore fire regimes and native grasses to mountainsides stripped of their small growth by overgrazing and their large growth by promiscuous lumbering. The New Dealers *withdrew* millions of acres from cultivation, acres that the nation, lacking an environmental policy respectful of the land, had permitted to be desolated. Much of the high dry west was and is not fit for the plow, nor is it fit for some kinds of grazing animals. Much of it has been *withheld* from cultivation or grazing. As it was, the New Deal came to the high plains after they had been despoiled of their natural grasses, and exposed to the wind. Shelterbelts were built after the fact, to replace the shelter that had been ripped away.

Did all the New Deal experiments in conservation, including shelterbelts, work to avert the desolating winds of the Dust Bowl in the next terrible years? In many places the sod was gone and it was too late. But not too late to teach a lesson—a lesson poignantly before us again today—that remediation is more costly than prevention. That is the lesson of the New Deal for us—some implications of which will emerge as we go forward. The history of the early 1930s reinforced the lesson that humans are part of nature and cannot safely exceed its limits.

As the first withdrawals of land were proving good for both land and farm prices, the great drought of the 1930s produced the Dust Bowl. This was not the first time in the memory of living persons that the natural West had risen in revolt. This was the second *natural* sagebrush rebellion—we'll turn back to the first, the early 1890s, in a moment. The lesson was the same then—we must do all we can to avoid imposing land uses upon terrain that cannot bear them, or nature will react catastrophically. If the land use has been destructive, catastrophes will come, raising the cost of restoration and forcing retreats from the angry land, like those of the 1890s and the 1930s. When people will not retreat, they will be driven off by nature itself, as the Okies were driven off.

Before we turn back to the 1890s, let us look unblinkingly at the 1990s— and the last two years. There is an eerie and instructive resemblance between the Dorothea Lange photographs of the shanty towns of the Okies and new pictures, taken this year, of tent and shack cities of refugees from fire and flood in Arizona and Missouri. We humans are slow to learn. We require several iterations to learn harsh lessons. In the 1920s and 1930s the abuse of the land came from agriculture. In the 1990s and in our time it has come from exurban development. In the 1930s the catastrophes were made of wind and earth—dust storms on a continental scale. In our time the catastrophes come from wind and fire—firestorms that have only taken forests and urban fringes so far, but one day soon will take whole towns.

The human consequences are the same. "As human beings" we are "faced with a condition and not a theory." We must not think of property, indeed we must, but we must put people first.

*   *   *

Well, there were the 1890s—but there was a lot of history before the 1890s. Franklin Roosevelt could draw wisdom from Grover Cleveland—and the 1890s, but we know some things Roosevelt could not know. Despite the achievements of Works Progress Administration (WPA) archaeology, Roosevelt did not have available to him some important facts about the longer experience of human beings on this continent, lessons imparted to us by the 60 years of archaeology and environmental experience, since the New Deal. Because we have learned more about the long cycles of human use and abuse of land and limits, we can set the New Deal and ourselves in a larger and longer cyclical context.

We humans have been around here a very long time, enjoying periods of construction and of high expectations, followed by other periods of destruction, decay, and disappointment, followed by renewal, reaffirmation, and rebuilding. The archaeological and architectural history of North America shows that we have repeatedly exceeded the limits set for us by nature and that nature has repeatedly taught us harsh lessons. Ruins are reminders that we have known on this continent recurring cycles of achievement and degradation. We have cultivated land better left alone. We have developed beyond natural limits. We have retreated in defeat and despair. But we have learned to recover and try in better places—where we have given ourselves and nature New Deals.

The record shows the first monumental architecture in North America to have been laid up earlier than the first stepped pyramids in Egypt. Five thousand years ago, the people of north-central Louisiana, at Frenchman's Bend, Watson Brake, Stelly, and Monte Sano, built great earthen buildings. Then something went wrong, and nothing so grand was tried again for 2,000 years, until in the time of Homer's Troy, about 1200 BC, the people of Poverty Point laid up their monumental buildings 90 miles from Watson Brake. Again, something went wrong; they stopped building and making sculpture and exchanging gifts throughout the great valley, before another time of recovery ensued, and social and architectural energy flagged for a thousand years. Then people tried again, and there was a great flowering of sculpture and large-scale geometric architecture, as Hopewell engineers and astronomers in the Ohio Valley, during what Europeans came to call "Roman times," deployed millions of basket loads of earth to correlate their architecture with their observations of the progress of the moon and sun and stars in time and space. Their were very sophisticated structures, using octagons, squares, circles, and trapezoids as the Prairie School architects would use them again, in glass and steel, brick—and earth as well—in some of the same places nearly 2,000 years later. Then . . . once more, limits were ruptured and centuries passed before another set of urban clusters formed in the tenth century. One metropolis grew at the intersection of the Missouri, Mississippi, and Ohio. In the West at the time, huge masonry structures went up in Chaco Canyon—more impressive buildings than any palace being built nearly at the same time for Charlemagne. Soon afterward large-scale building began again in the Mississippi Valley. In the twelfth century, when Scandinavians knelt before their bishop on Greenland, the metropolis of Cahokia, opposite St. Louis, contained more people than either London or Rome. The Mississippians and the Anasazi flourished, and then, once more, exceeded their limits, and grand endeavors ceased. Not only that—even before European microbes and European conquistadors arrived, humans had retreated from large, previously cultivated and "developed" regions, as they retreated in the 1890s and the 1930s—and it is likely that we must soon retreat again. By the year 1400, the Vacant Quarter had opened where the

Cahokian metropolis had flourished in the Mississippi Valley. Chaco Canyon and Mesa Verde had been evacuated. When De Soto entered the valley of the Savannah River in 1540 he found that once densely settled region deserted—it was the "Desert of Ocute."

The ruins of human aspiration were everywhere. There was no charming pre-Columbian Eden, no primeval Arcadia—humans had many times stretched the capacity of the land and air and water to absorb them, and many times the survivors of Nature's retaliation had fled from the desolation left by excessive cultivation and development. Those survivors were vulnerable in 1492.

There was a crucial exception—the Pueblo communities of the Rio Grande Valley, where the Indians refused to die or to relinquish their traditional forms of building and worship—and where their own relationship to nature had come to assume a pattern of disciplined steely elasticity, skeptical but active. The Puebloans were the first exemplars of Smartgrowth—after repeated failures, their communities concentrated habitation and controlled the use of land, and water, and fire. They protected the people. For them, property came next—really came next. The Puebloans had learned not to sprawl—and that is why Oriabe and Acoma are the oldest continuously occupied towns in the United States. Smartgrowth is not a new idea—after the Pueblos it reappeared in the Garden Cities component of New Deal planning.

Traditional wisdom recognizes limits. Twenty thousand years of experience admonishes us to repent in time our contamination of land with toxic waste, our polluting the waters with acid rain, our ravaging the plains into caliche wastelands and dunes of dust. We are reminded of this today by the smoke and ash of fires made catastrophic by insufficient diffidence—insufficient self-skepticism—the face of natural processes and natural limits.

*   *   *

This chronicle urges upon us a turn in our discussion from nature back to people, to specific people, and to Franklin Roosevelt in particular—the cheerful redeemer. From whence did his cheerfulness come? It came, I think, from another expression of his consciousness of time. Not only did time, in history, force upon him "a wariness of unforeseen consequences—and beyond that, to a recognition that there are four dimensions in the real world—that time passes and changes everything you thought you knew." His consciousness of the dimension of time permitted him to feel a partnership with people he respected in other eras. He could remain so resilient, so persistent because he had company in serving the creed of *steely elasticity,* and redemptive skepticism with other members of what Emerson called "the *party* of hope."

As a historically minded man, Roosevelt was temperamentally a Founding Father. The Founding Fathers were not only his predecessors but also his partners. Like his hero, Lincoln, they evoked hope in crisis—summoning from their contemporaries a recognition of the possibilities of crisis. The Founders—Roosevelt's partners, and our partners—had to be hopeful, to believe in redemption, for they had to survive an eight-year civil war, and it must be recalled, the failure of their first experiment in self-government, that of the Confederation. The New Deal was also built upon the ruins of an earlier failure, in the ashes of the first depression. It required courage equal to that of the Founders, for though they suffered disappointment in their expectations of each other, they did not face an assault from nature as well like the dust storms of the great drought, nor did they experience economic collapse like the second great depression, that of 1937.

When we look at the Great Seal of the United States inscribed on our dollar bill, it could just as well be the emblem of the New Deal. The Founders said they meant to establish a "Novo Ordo Seclorum." The phrase had two possible translations, the first, "A New Order in the Universe," situated their endeavor in *space*—in geography, and by implication in that portion of the universe they occupied, the American land. The second placed them in *time* as "A New Order among the Ages." Thus the Founders blazoned upon their seal and their currency, as if upon their shields, an intention to initiate an *epoch,* morally and intellectually, within a *place in geography* in which a new kind of human might be nurtured. Thus an environmental consciousness was there from the beginning.

Let's be sure we see those Founders as they were, people much like us. Yes, they were only "respectable lawyers, doctors, merchants, and landowners," in the faintly condescending description of Winston Churchill. Yes, they may have seemed at the outset "nervous at the onrush of events" just as many of Franklin Roosevelt's associates were nervous, and should have been. Yet, drab in their dun-colored bourgeois clothing as they might have appeared to gorgeous courtiers flitting above the decay of the old order in Europe, the Founders were determined to transform the world they found. As Jefferson wrote Joseph Priestley, "we can no longer say there is nothing new under the sun. For this whole chapter in the history of man is new." In his old age, Jefferson reminded his colleagues of that glorious explosion of possibilities when "our Revolution . . . presented us an album on which we were free to write what we pleased."

That album *was* geographic as well as political. Jefferson put the Mississippi River in its centerfold with the Louisiana Purchase. Many of the hopes of the Founders were disappointed. One of those hopes, about which I have written a book, *Mr. Jefferson's Lost Cause,* was that the great geographic prize of the revolution, a third of a continent, would be allocated for the use of family farmers. It was believed by many of the Founders that working land is good for people. Yeomen became "the chosen people of God" because of the benign effects upon them of working in and with the land. The republic had an interest in the yeoman's presence on the land because agriculture was good for him and he would, as a consequence, be good for the nation. This therapeutic view of land won its proudest victory over a purely commodity theory of land in the Homestead Act.

When Jefferson dignified their work, 90 percent or more of Americans "labored in the earth." Jefferson's phrases rang in the rhetoric of William Jennings Bryan in the 1890s, when 30 percent still farmed. When the same phrases were uttered by Robert Lafollette in the 1920s, 17 percent of Americans gave their profession as "farmer." Only 5 percent were full-time farmers when Hubert Humphrey brought the litany forth again in his 1968 campaign. Counting real farmers is not easy these days, but the percentage is probably down below four, today. Does that render invalid the idea that humans are part of nature? Need we now be embarrassed to assert that even we, urbanized, motorized folk, cannot separate ourselves from the earth and survive? I do not think so. Nor do I believe it sentimental for us to affirm the essential meaning of Franklin Roosevelt's reiteration of Jeffersonian agrarianism in his Farm Speech of 1937:

> Sturdy rural institutions beget self-reliance and independence of judgement. Sickly rural institutions beget dependency and incapacity to bear the responsibilities of citizenship . . . .[On the family farmer] continuance of the democratic process in this country to no small extent depends.[4]

Roosevelt's Secretary of Agriculture Claude Wickard returned to Monticello in 1944 to assert that he and his president endorsed Jefferson's precept that the "peculiar deposit . . . of the Almighty's virtue" had been placed with "those who labored in the earth . . . on their own soil . . . [with] their own industry." How do we respond, now that "the family-sized farm," "the seat of liberty," is itself an endangered social species? [5]

Though the proportion of our population who labor "in the earth . . . on their own soil" has declined, the essence of Jefferson's and Roosevelt's agrarianism—situating humans as parts of nature—remains valid. Humans are part of nature. We cannot escape the consequences of what we do to nature, nor can we function fully without regular exposure to earthy reality, with the irresistible truths of the seasons, with the elements of air, fire, earth, and water. However we may hide behind office walls or be distracted from large facts by the computations of small ones, however we may seek bemusement in computerized virtuality—we are parts of nature.

During the New Deal years, that fact was expressed in wilderness protection and in active intervention in land-use policies both in federal lands and through agriculture policy, and on private lands. Preservation and intervention are the two faces of active conservation, expressed since the New Deal in the Wilderness Act and the Endangered Species Act, the Clean Water Act and the Clean Air Act. They were expressed nearly a hundred years before the New Deal in the 1840s, when George Perkins Marsh was in Congress while Henry David Thoreau was at Walden Pond. Jefferson had said that in family farming was the salvation of the republic; Thoreau said that "in wildness is the preservation of the world"; Marsh anticipated Wendell Berry's dictum that "in human culture is the preservation of wildness." Marsh was a practical fellow—his great work was entitled *The Earth as Modified by Human Action.* While Thoreau was escaping to nature, Marsh was urging that we permit some part of nature to escape from us.

Let's start with Thoreau. His interest, like Jefferson's, was in the therapeutic value of land, in what nature does for us. If working the farm was good for people, going to wilderness was better. There should be some place where we would abstain from working. The preservation of wilderness is the ultimate act of skepticism—self-skepticism. Wilderness is that which lies beyond our anxious self-assertion. It is the proximate metaphor for that wide universe which, when we pray, we acknowledge to be even beyond our ken. We humans are part of nature, so our wilderness legislation is an acknowledgment of our delinquencies elsewhere, especially our transgressions of nature's limits.

Wilderness preservation was a central aspect of New Deal environmental policy, well before the Wilderness Act. And there is no doubt in my mind that Franklin Roosevelt, as a religious man, thought of the universe as intentional, as a creation—not necessarily all at once, nor necessarily taking only a week's time—but intentional. As such, he and Harold Ickes and Henry Wallace treated all its parts as valuable—all its species, all its mountains, waters, fields, and oceans. Agricultural reform, and the withdrawal acreage, the restorative work of the Civilian Conservation Corps (CCC) and the WPA expressed that veneration.

Though we do not conventionally genuflect as we pass the sign, "wilderness area," it would not be odd if we were to do so. Wilderness is a place and it is also an enigma, a profound puzzle—a place in which the mystery of life shows itself in its perpetually changing, infinitely various affirmations. Our proud, willful, often foolish and heedless species is learning all the time how little we really *do* know, how little we *do* control. All the essentials of life, including birth, love, and death, are intrusions by the unknown and by the essentially unpredictable into our well-planned, scrupulously managed, manicured lives.

Wilderness areas are not big zoos; *we* are in the zoos, *they* are outside the zoos. That is why most of the great national parks and monuments of the New Deal era—Big Bend, Shenandoah, Joshua Tree, North Cascades, Olympic, Great Smokies, Everglades, and Kings Canyon, were big enough to permit the full range of life within them. I hope that secularists among you will not be distressed if I suggest that wilderness is a sort of physical, geographical sabbath. In wilderness we can find surcease from the consequences of our own bad management elsewhere, of what we have done to the world and to ourselves "during the rest of the week," so to speak. Harold Ickes was the most important politician ever to take the cause of wilderness seriously, and work at it, in the tradition of Thoreau.

New Dealers dealt with people living outside wilderness. They, like us, lived with the consequences of transgressions of natural limits. They, like us, were determined to prevent further transgression—owing as much to the tradition of George Perkins Marsh as the tradition of Thoreau. Thoreau transcended. Marsh interceded. More than Thoreau or John Muir or Ickes, Marsh required his fellow countrymen to pay attention to the lessons of human depredation outside wilderness, outside parks, outside zoos and botanical gardens. Marsh asserted that the land—all the land—has rights, the further abuse of which could make of his country a wasteland as squalid as the once Fertile Crescent, or Sicily, or, had he been reading Francis Bacon, Ireland as well. The reformers then and later took up environmentalism as they took up women's rights, abolitionism, and civil rights. Their opponents were heedlessness, selfishness, arrogance—what is now called "dominance."

Yet after the 1840s of Thoreau and Marsh the fervor abated. The legions of selfishness and arrogance, and the battalions enrolled under the banner of King Midas, were still mighty. The 1850s were like the 1790s. Though the rest of the agricultural economy collapsed into the farm depression of 1857, the plantation South enjoyed its final prosperous paroxysm, devastating hundreds of thousands more acres by a system of unrepentant exploitation of the land and labor force alike. The planters got very rich and very proud, threatened secession, and when the rest of the nation elected Abraham Lincoln, finally seceded. During that catastrophic conflict we call our Civil War, Lincoln found time to proclaim an environmental policy. The Homestead Act was finally passed because the planters' senators could no longer block it. It was late—much of the best agricultural land in the nation was settled without it—but even for the tallgrass prairie states it offered a more ecologically responsible system of agriculture than plantation slavery. As the New Deal later made articulate, agricultural policy is always a part of environmental policy.

Lincoln also gave the first impulse to the National Park System by insisting that Yosemite be preserved from exploitation. In that unacknowledged way he provided to other species—beyond our own—their own "new birth of freedom."

There is in Lincoln's way of dealing with the chief problem of his time—slavery—instruction for how we should deal with another terrible legacy—the treatment by our species of the other species of this American land. You recall: "Fondly do we hope, fervently do we pray—that this mighty scourge of war may speedily pass away." Fondly we may hope that we will not kill more of our fellow creatures—on this continent, and where a "New Order"—Novo Ordo—is necessary. You recall as well Lincoln's next words: "Yet if God wills that it continue, until all the wealth piled by the bond-man's two-hundred and fifty years of unrequited toil shall be sunk, and every drop of blood drawn with the lash, shall be paid by another drawn by the sword, as it was said three thousand years ago, and so it must be said 'the judgements of the lord are true and righteous altogether.' "

As we look to western horizons aflame with the consequences of our accumulated transgressions upon the natural order, let us hope for mercy as well as righteousness. Lincoln was attacked for that speech—he wrote just afterward that he knew why it was "not immediately popular. Men are not flattered by being shown that there has been a difference of opinion between the Almighty and them. To deny it," he wrote, "in this case, is to deny that there is a God governing the world. It is a truth which I thought needed to be told."

In our case, to deny that we have transgressed our limits would be to refuse to heed the lessons of fire and flood. Franklin Roosevelt did not deny the state of the American earth, nor did he refuse to heed the lessons of dust and degradation. Roosevelt quoted Lincoln as having said: "we must candidly avow that none of us has done all that we might, and, also, that it is never too late for any of us to do better." That is the spirit of the New Deal of the 1930s, and it must be the spirit of a New Deal for Nature in our own time. After Lincoln's death, Frederick Douglass summarized Lincoln's legacy—it was refounding of the American Republic. "Lincoln," wrote Douglas, "believed that his first responsibility was to preserve the Union because that Union itself had a redemptive power." A redemptive power! A power to redeem the Founders' hope, of a New Order—and our hopes, periodically, whenever our nation falls from grace. It may do so—and climb back out again—by a number of routes.

After the Civil War, slavery was abolished. More civil wars with the Indians brought the western plains into the white farmers' and grazers' domain, and a run of wet years induced a western migration into every higher and dryer country. Nature responded with the droughts and blizzards of the late 1880s—and now the 1890s—in the 1890s came the first great retreat from the West. Convoys of prairie schooners headed back eastward. The farmers blamed the bankers, the railroads, and the Gold Standard, but the real story was that the limits had been violated. "Fully half the people of western Kansas left the country between 1888 and 1892"—in the farmers' first great retreat from the West. Twenty-six counties of South Dakota emptied out. In 1891, 18,000 prairie schooners crossed the Missouri from Nebraska, going east. And what did they leave behind? A landscape as ravaged as the loess hills of Mississippi and the piedmont of Georgia after the plantation system had done its antebellum worst.[6]

The farmers remembered, but the rest of the nation settled into the next normalcy—the 1890s, Big Business supremacy, White supremacy, Jim Crow, and an illusion of human supremacy over nature. The Cooperative Movement was launched. But it did not include cooperation between humans and other species. As the century turned, the Progressive Era flowered in the compost of Populist exertion. The Forest Service was founded in 1905, under Gifford Pinchot.

Then, in the first years of the world war in Europe, and of another transient and unsustainable rainy period, Europe required and received an immense quantity of American farm products. Drawn by Europe's desperate demand, the farmers pressed again across the natural limits—steel plows broke the high plains, people talked of "dust mulch" while the wind was down, and nature commenced preparation of its counterattack.

The war came to an end. Demand collapsed. Farm income dropped from 17 billion in 1919 to 10.5 million in 1921. The Great Depression came early for the farmers—many of whom responded by borrowing more, plowing more, grazing more, breaching the limits ever more deeply, and becoming poorer and poorer as they were trapped in their own productivity. Calvin Coolidge responded by opining that "the farmers have never made money. I don't believe we can do much about it." There was no more sense of an environmental disaster in the making on the high–dry agricultural frontier than there is today on the high–dry urban and exurban frontier.

No leader of government in the 1920s sought to "do much" as one quarter of the nation went on a stock-buying binge while three quarters of its families made a go of less than $3,000 a year—the Jazz Age for some and the Blues for many. The environmental depression ensued from the farm depression, and the nation danced the Charleston. The New Deal got started after farm income, having caught up a little, fell in the early 1930s by another 30 percent.

In the 1880s, nature had retaliated with drought and blizzards, but until the 1920s most of the grasslands were still intact in the West, and most of the overgrazed mountain slopes were uninhabited. The New Deal was having some success with withdrawing acreage from production when nature grew impatient and withdrew the topsoil. The great drought ensued, and then the great blowaway. Between 1931 and 1934 the Dust Bowl blew the exposed surface of the plains into the mouths and ears and nostrils of an already defeated people. Many died of suffocation. Cattle died. Crops were completely covered with dust-dunes. The sky in Chicago was so dusty that there was twilight all day—for weeks on end. Dust from Kansas blew into Philadelphia and New York. The Grapes of Wrath. Steinbeck. Okies. Dorothea Lange. Margaret Burke White.

It seems so distant now, doesn't it—until we look at the sky above Denver and Boulder and Albuquerque—it isn't dust, now, it's ashes. The sky is yellow, today, over much of the West, as it was in 1933. Nature is sending winds to drive fires over overgrazed and overtimbered slopes as nature sent winds to drive the vulnerable surface from overplowed and overgrazed plains. The lesson is the same—we have transgressed our limits—but now we have added a further transgression—we have situated ourselves where we once only situated our cows. In the 1880s, 1920s, and 1930s, herds and farms were pressed beyond the legitimate limits of the human domain in the West. In the 1990s our exurban developments did so. Once illegitimate agriculture affronted nature. Now we provoke nature again, but with illegitimate development.

Dr. New Deal was replaced on the case by Dr. Win-the-War, and it is a tribute to leaders of the probity and wisdom of Dwight David Eisenhower and George Marshall that we did not follow World War II with another cycle of heedlessness, corruption, or post-war depression, and affected a decent transition to the 1960s and 1970s, when we returned yet once more to the spirit of the 1770s, of the 1840s, of the Populists and Progressives, and of the New Deal. Even in this less heady period, we may recall our fifth redemptive epoch, in which Civil Rights were once more linked to environmental responsibility, and anticipate cheerfully the sixth.

The 1990s have repeated the 1920s—and the 1850s, and the 1870s. We are ready for a rebirth of the better America—for a New Deal for Nature and Nature's People.

We live in a time in which another great withdrawal must occur. Our intrusions into fire-prone canyons and flood-prone wetlands are as dangerous to people as was breaking the sod and ripping away the grass of the high plains. We must, over time, withdraw from fire-prone canyons and flood-prone wetlands and hurricane-prone shores. We must retreat from those perilous places—or, at the very least, refrain from further invasion of them. Retreat—another retreat like those of the 1890s and 1930s, will be painful and expensive. But it will be less painful and less expensive than perpetual rescue and disaster relief—and degradation of the land. We have already retired roughly 10 percent of our cropland in the late 1980s under a Conservation Reserve Program, paying farmers to take eroding land out of production. People don't get killed fighting erosion—but they do get killed fighting fires. We must retire from development of fire-prone land, as well.

Sooner or later we will come to it. Sooner we hope, because if later many more young men and women, fire fighters and flood fighters will die as we are too slow in learning. We must use every device of sophisticated economics to diminish the incentives that have driven people to settle most heavily in the last decade in the driest of our states—and the most dangerous. Mortgages, insurance, and reinsurance, power-line construction and road building—all these devices of policy were used by the New Deal for benign purposes. And all these devices have been used in recent decades to advance the interests of developers and to put people into danger.

Finally, and most importantly, we must reenroll a CCC, on a much larger scale than Americorps, to clean up our mess—our tinder, accumulated over years of fire suppression, our ravaged slopes, and our filled and contaminated bosks. We must free our rivers and free small natural fires. It's time for a New Deal for Nature—and for Humans—for Nature's People.

NOTES

1. Stories and quotations from James McG. Burns, *Roosevelt, The Lion and the Fox* (New York: Harcourt Brace Jovanovich, 1956) 245–46.
2. Ibid.
3. Ibid.
4. Roosevelt quoted in Alfred Whitney Griswold, *Farming* 15.
5. Wickard quoted in *Griswold and Democracy* (New Haven CT: Yale University Press, 1943) 18.
6. Data and quotation from John D. Hicks, *The Populist Revolt: A History of the Farmer's Alliance & the People's Party* (Nebraska: University of Nebraska Press, 1961) 32.

Notes on Contributors

Richard N.L. Andrews is Professor of Public Policy, University of North Carolina, Chapel Hill. Professor Andrews has published numerous articles and books on environmental history and policy, including: *Managing the Environment, Managing Ourselves: A History of American Environmental Policy* (Yale University Press, 1999) and *Environmental Policy and Administrative Change* (Lexington Books, 1976).

Brian Black is Professor of History and Environmental Studies, Pennsylvania State University at Altoona. He is the author of numerous publications on history, energy, and the environment, including: *Petrolia: The Landscape of America's First Oil Boom* (Johns Hopkins Press, 2000). He is currently working on a study of the history of preservation efforts at the Gettysburg Battlefield.

Henry L. Henderson is president of Policy Solutions Limited, an environmental consulting firm based in Chicago, and is Senior Lecturer in Environmental Studies at the University of Chicago. He served as the first Commissioner of Environment for the City of Chicago and Assistant Attorney General for the State of Illinois, specializing in environmental law. He is Chairman of the Chicago Environmental Fund.

Roger G. Kennedy is a distinguished author, public leader, and civic activist. From 1993 to 1997, Mr. Kennedy served as the fourteenth Director of the United States National Park Service. Prior to that, he served as the Director of the National Museums of American History, Smithsonian Institution, Vice President of the Ford Foundation, and as Special Assistant to the Attorney General and the Secretaries of the Department of Health, Education & Welfare, and Department of Labor. Mr. Kennedy is the author of numerous articles and books on environment, politics, architecture, and history, including: *Burr, Hamilton and Jefferson, Rediscovering America, Figures in a Moving Landscape, and Mr. Jefferson's Lost Cause*.

John Leshy is the Harry D. Sunderland distinguished Professor of Property Law Hastings College of Law in San Francisco at the University of California. Professor Leshy served in the United States Department of Justice, as Solicitor of the U.S. Department of the Interior, and as Associate Solicitor for Energy and Resources of the Department of Interior. He has also served with the Natural Resources Defense Council. Professor Leshy has published numerous articles and books on environmental law and policy, including *Federal Public Land and Resources Law*, with George

Cameron Coggins and Charles Wilkinson (4th ed., Foundation Press, 2001), and *Legal Control of Water Resources,* with Joseph Sax, Robert Abrams, and Barton Thompson (3rd ed., West Group, 2001).

Jim Lyons is Professor in the Practice of Resource Management, Yale University. He served for eight years as Under Secretary for Natural Resources and Environment in the Department of Agriculture, with oversight responsibility for the Forest Service and the Natural Resources Conservation Service.

Neil Maher is an assistant professor in the Federated History Department at the New Jersey Institute of Technology and Rutgers University, Newark, where he teaches American environmental history, urban history, and the history of technology and health. This essay is part of his book manuscript titled, *Nature's New Deal: The Civilian Conservation Corps and the Roots of the American Environmental Movement, 1929–1942.*

Sarah T. Phillips, Assistant Professor of History at Columbia University, specializes in twentieth-century American politics and rural policy and has published articles in Agricultural History and Environmental History. The material presented in this collection is excerpted from her book, *This Land, This Nation: Conservation, Rural America, and the New Deal,* forthcoming from Cambridge University Press.

John F. Sears is Associate Editor of the Eleanor Roosevelt Papers and an independent scholar. From 1986 until 1999, sears served as Executive Director of the Franklin and Eleanor Roosevelt Institute in Hyde Park, New York where, among other projects, he supervised the restoration of FDR's Top Cottage, and helped establish the Roosevelt Foundation for United States Studies at Moscow State University. Sears is the author of *Sacred Places: American Tourist Attractions in the Nineteenth Century,* editor of the Penguin classic edition of Henry James's *American Scene* and of *FDR and the Future of Liberalism,* and coeditor of *FDR and His Contemporaries.* He lives in the town of Hawley, Massachusetts where he owns a Tree farm.

Paul Sutter is Professor of History at the University of Georgia, Athens Georgia. He is the author of numerous publications on environmental policy and history, including: *Driven Wild: How the Fight Against the Automobile Launched the Modern Wilderness Movement* (University of Washington Press, 2002), and "Terra Incognita: The Neglected History of Interwar Environmental Thought and Politics," *Reviews in American History,* 29 (June 2001).

A. Dan Tarlock is Professor of Law and Codirector of the Program on Environmental and Energy Law at Chicago Kent College of Law. He serves on numerous environmental commissions on land use, water policy, and resource protection, such as the Western Water Policy Review Committee. He is the author of numerous articles and publications on environmental Law and Policy, including *Environmental Protection: Law and Policy (Aspen Publishing 2003), Law of Water Rights and Resources* (Clark Boardman Callaghan, 1988, 1998), four legal casebooks, and seminal articles such as "The Non-Equilibrium Paradigm in Ecology and the Partial Unraveling of Environmental Law," 27 *Loyola of Los Angeles Law Review* 1121 (1994).

David B. Woolner is Assistant Professor of History and Political Science at Marist College in Poughkeepsie, New York, Executive Director of the Franklin and Eleanor Roosevelt Institute in Hyde Park, New York, and a member of the Board of Directors of Hudson River Heritage. He is the author of *Cordell Hull, Anthony Eden and the Search for Anglo-American Cooperation 1933–1938* (forthcoming from Praeger Press) and the editor of *The Second Quebec Conference Revisited: Waging War, Formulating Peace; Canada, Great Britain, and The United States in 1944–45* (Palgrave/St. Martin's Press, 1998), and *FDR, the Vatican, and the Roman Catholic Church in America 1933–1945* (Palgrave, 2003).

INDEX